Lars Döhmann / Norbert Schäfer

PORSCHE INSIGHT

Lars Döhmann / Norbert Schäfer

PORSCHE INSIGHT

The technical illustrations since 1975

WHITE STAR PUBLISHERS

Contents

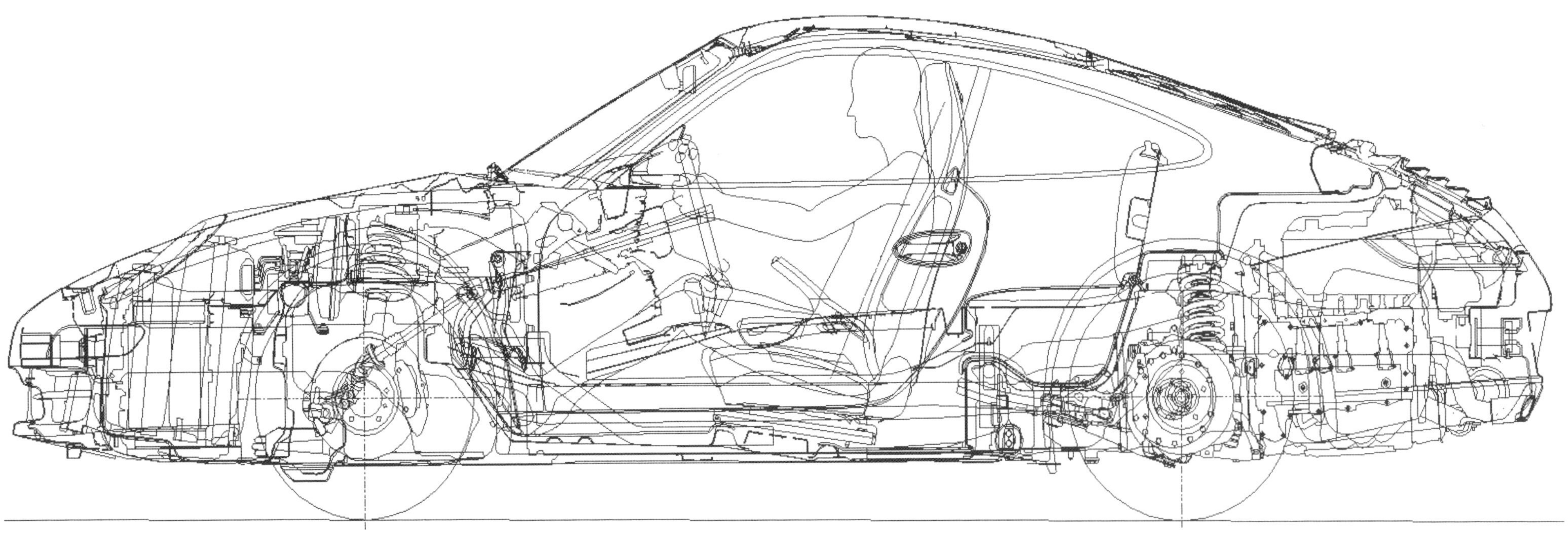

About this book
Insight in depth into Porsche engineering 8

Technical Art
Over 30 years of technical drawing 10

Racing cars 19

Porsche 935
A success with a flat nose 20

Porsche 936
A Spyder that won Le Mans three times 26

Classics 33

Porsche 924
An ever-increasing appreciation 34

Porsche 944
The fastest four-cylinder cabriolet in the world 40

Porsche 968
The crowning of the four-cylinder front-engine series 50

Porsche 928
The V8-GT with transaxle 56

Porsche Carrera GT
An uncompromising symbol of technology 70

Porsche 911
A development that would last for decades 80
Normal and turbo engines 82
964 – Fundamental revision 88
993 – The last air-cooled engine 96
996 – The novelty of water-cooling 102
997 – Forever young 114
GT2 – For competition and road use 130
GT3 – Inspired by motorsport 138

Modern models 146

Porsche Boxster
A successful mid-engine roadster 148

The new Boxster
Mid-engine roadster by Porsche 158

Porsche Cayman
The racy coupé 162

Porsche Cayenne
The multi-talented sports vehicle 168

Panamera GTS
Sports car with four doors 182

Package solution 186

Car concept and package 188
Porsche specification 192
Bibliography, acknowledgments 198

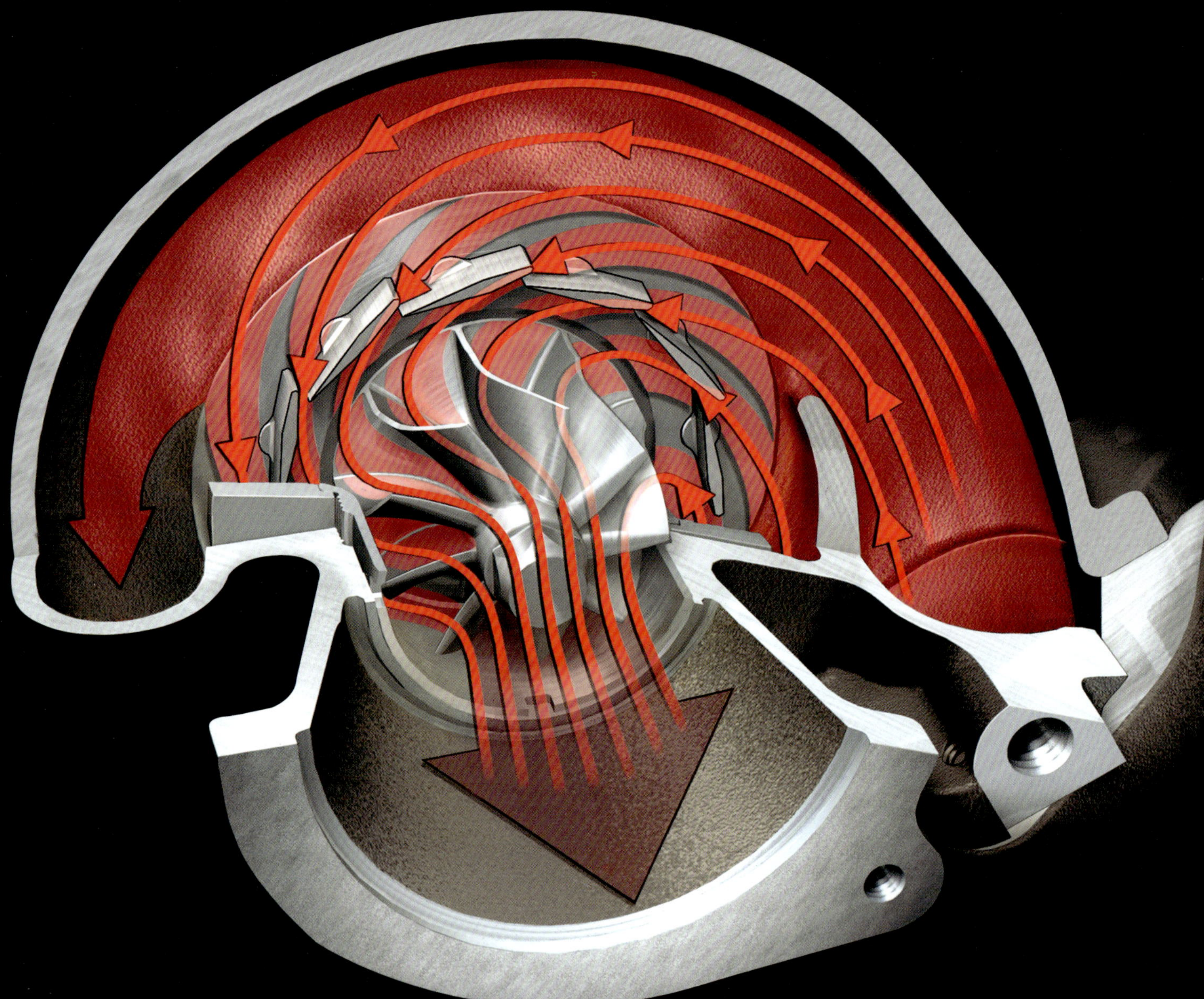

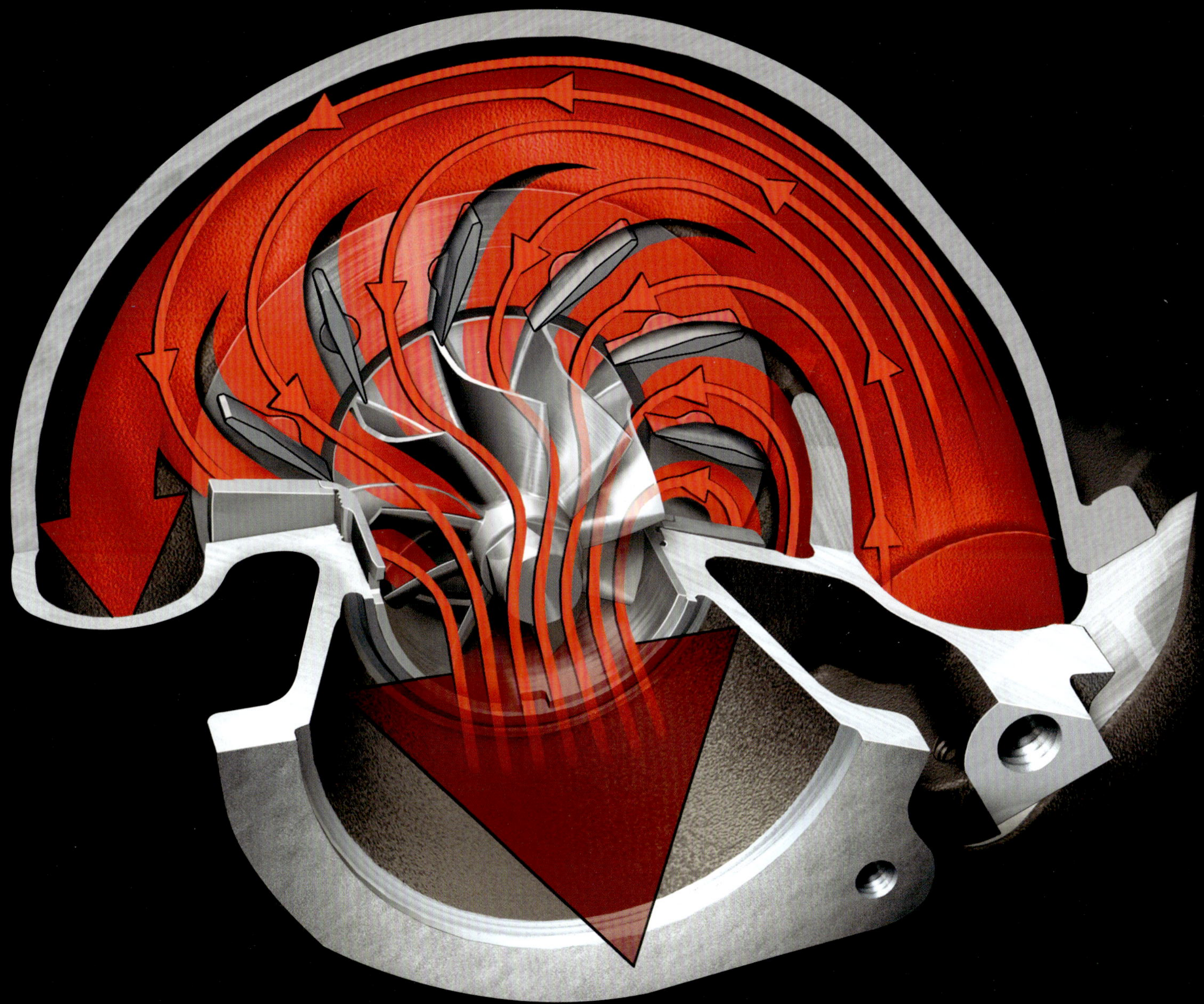

About this book

Insight in depth into Porsche engineering

» The name Porsche is closely associated with the development of mass motoring through the Volkswagen Beetle. This also explains why sports cars, such as the legendary Type 64 Berlin-Rom-Wagen, were built with parts of the Beetle that itself was still in a phase of development. The first Porsche saw the light in 1948 in Gmünd in Austria, and it could only be built because low-cost mass-production technology was available to build a special car.

The prototype with a mid-engine and tubular chassis, the Porsche 356, became a lasting success that was then slowly and meticulously developed further. The 901/911 was a further icon that would fascinate anyone interested in cars, and this was certainly true in my case. As the editor of *Motor Klassik*, in March 1988 I was asked to write a cover story, and so was able to push the prototype 754 around in the photographic studio. Three or four years later, as test driver for the magazine, I drove a sedan at a speed of 174 mph (280 km/h) on the expressway for the first time: my hands, clammy with sweat, were clenched to the steering wheel of a Porsche 911. A Carrera 4 Cabriolet was also once involved. To the question I am often asked, what car would I choose if I had a completely free choice, I always reply: the maker, Porsche, and the model, the 911.

A good three years later, after many endurance races in a rear-engined VW Beetle Fun Cup racer, I was able to purchase a 1979 911 SC Targa. It had meanwhile been legitimized with the OldTimer designation, and because of my vast experience in the field of driving dynamics, the palms of my hands were no longer clammy. But my fascination for that unique maker, and that legendary model, had remained unchanged. The workshop manual is always on the shelf ready to hand because the technology of an OldTimer always needs a little attention.

So for me it was absolutely fantastic when I was commissioned to write this book in spring 2010. In particular, cutaway drawings had long been among the pictures that I had studied in detail through a magnifying glass. They are easier to understand than long verbal descriptions of how the complex assembly of a car operates. This is why the texts accompanying the illustrations by the talented graphic artist Norbert Schäfer and his colleagues at the Technical Art company have been kept concise, while still reflecting the numerous technical developments. But the text cannot, and does not, provide complete technical descriptions: there are other books that do this. In this book more than most, it is evident that a picture is worth more than a thousand words. Each of these extremely precise, detailed illustrations is the result of several weeks' work. They are absolutely true to scale, so that they can even be used in restoration to ensure the exact reproduction of the original.

Besides the technical development of the model ranges included in the book, you can also follow the development in the techniques used to make the illustrations. Between 1975, when the first drawings in this book were put down on paper, and the year 2008, the manual work carried out with pencil, brush, and paint has been replaced by the computer. Sophisticated data processing capabilities and CAD (computer-aided design) have made this possible.

I hope you will enjoy reading and, even more, studying this unique volume.

Lars Döhmann
Neustadt, Germany, July 2010

Technical Art

Over 30 years of technical drawing

›› To transfer onto paper the ideas and illustrations of the best designers and engineers! This was the great challenge Ludwig Eberl, and Norbert Schäfer set themselves in the early years of their company Technical Art, working in the same way as the old masters: with brush, paint, and skill. Even when the tools in their joint field of activity changed radically with the introduction of computer graphics and digital image processing, some things remained unchanged: their fascination with the beauty of technology, their love of detail, and their artistic passion for their work.

» The history of graphic illustration, and therefore also the history of Technical Art, reflects the impressive development from traditional manual work to the most up-to-date, multi-media imaging techniques.

Already before the creation of Technical Art in 1975, Ludwig Eberl, born in 1929 (died 1995), and Norbert Schäfer, born in 1940, were working together in their well-known retouching studio. Because the two had also developed outstanding skills as illustrators, their new graphic art studio soon attracted the attention of several leading automobile manufacturers. It was obvious that the two graphic artists not only understood the language of designers but also that they had a particular passion for the representation of technical details. Thus, many high quality illustrations were published under the name Technical Art that received international recognition after their publication in the press all over the world.

As more of their illustrations were published, so the illustrators themselves became known among the community of experts, and the two German illustrators were even mentioned in the Japanese press. Their reputation grew so that after a few years Technical Art became the first port of call for many leading industrial companies, and this is still the case today. Technical Art's special collaboration with Porsche started as early as 1975, and is still going strong today, sufficiently so to fill an entire book.

The longstanding collaboration with Porsche is all the more remarkable when one bears in mind not only the incredible pace of technological change and the complexity of vehicle design over the period, but also of the changes in visualization techniques that have taken place during the three decades of the partnership. Although centuries have elapsed since the invention of the ink pen and crayons, today barely six months will go by without Technical Art using a new generation of three-dimensional image-processing software.

The company's remarkably large archives include many drawings of vehicles of all kinds dating from the first nineteen years of Technical Art that were still produced using conventional illustration techniques. Anyone allowed to forage in these archives will find some fascinating competition cars, fast Formula 1 single-seaters, sleek motorbikes, and of course also a whole range of mass-produced cars that one has seen on the roads, all drawn by hand. What still stands out in these illustrations today is the astounding wealth of detail showing the technical functions of every engine component, down to the last cable connection.

This conventional method used in technical illustrations was divided into several stages. At first as much photographic information as possible about forms and materials would be collected. After a period of research and some initial sketches, a design drawing in perspective was drawn on tracing paper that was mounted a drawing board. The third step involved the time-consuming coloring of the illustration, much of this being carried out using brushes and then refined with an airbrush. This required not only a technical understanding of how to assemble the individual fragments of information available but also the creative ability to present all the components accurately with the correct color and reflectance.

At the time the offices of Technical Art looked rather like a painter's studio. Then at the beginning of 1993 this changed with the introduction of the then most powerful computer graphics system in the world, the Quantel Paintbox. Admittedly, the classic set of illustrator's tools did not disappear immediately from the designers' desks, but gradually the computer was used increasingly often in several important stages in the design. The great advantage was not just the modern look of the images but the ability to work in a modular way. Complete illustrations were no longer drawn on sheets of paper: first, the individual components of the future illustration would be worked out, then they were assembled in the computer like the pieces of a puzzle to make an overall picture. In the 1990s, the conventional preliminary work on paper continued to play a fundamental part in the process. In a second stage, the image fragments drawn in this way were then scanned and thus converted into data, so that they could be assembled in the Quantel Paintbox and digitally processed. This had the undisputed advantage that they could react more quickly to clients' requirements or make technical changes to the model in a much shorter time, since the individual pieces of an image could be removed and replaced by new or modified detailed views. The new possibilities offered by this method of working were fully exploited for the first time in 1995 in creating the graphics of the Porsche C4 993 Turbo.

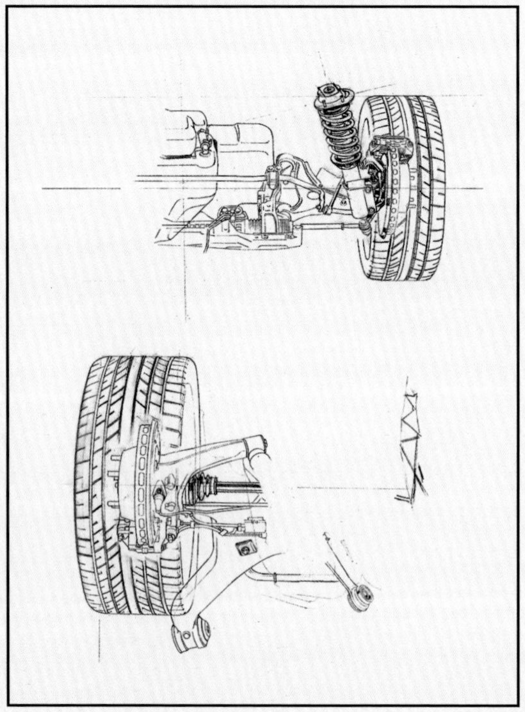

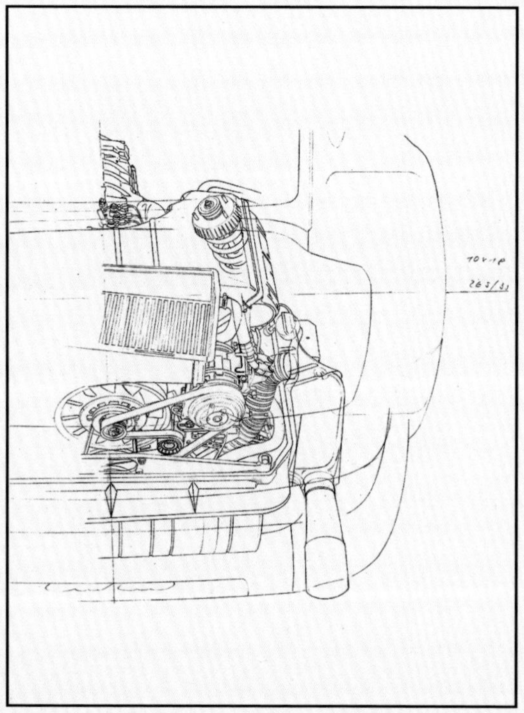

The graphic representation of the Carrera 996 was created in the same way in 1997 and 1998, with a complete series of images of variants being created for the first time. The advantage was that it was possible to provide very detailed, comprehensive documentation covering everything from the chassis to the highly complex overall technology. The C 996 was the first vehicle that was entirely processed using this extensive presentation, and this was so successful that Porsche also asked Technical Art to create a visualization of the Boxster.

In order to remain at the forefront of this fast-moving and constantly developing digital technology Technical Art was, shortly before the third millennium, among the first to venture into the three-dimensional world with the purchase of a Silicon Graphics Computer System. But it was only in the new millennium, with the Porsche Cayenne project, that the data for a new vehicle was entirely prepared in 3D, so that it became possible to create a visualization of a complete vehicle from these data, both as a whole and with a multitude of detailed drawings.

At the start photographic research for colors and the composition of materials was still necessary, but with the availability of 3D design data, this could be entered into the appropriate 3D software so that complete 3D processing of the images could take place in the computer. For the illustrator this involved constructing and selecting a scene, then materializing every single detail, including giving it the appropriate surface and texture. After this it can be highlighted and rendered, in other words computed. Only then can everything be assembled and reworked by the computer image-processing program. The Cayenne was the first representation of a complete vehicle known to us that was produced in this way. Now all this sounds much easier than it is in reality. Even when the digital construction plans have produced a brilliant model, it is still the illustrators who, admittedly using new digital tools but always with great talent and an eye for aesthetics, create realistic views of a vehicle. The potential of 3D technology has not only radically changed the working methods of Technical Art: at the same time, the requirements have also become much greater. Today the illustrators do not simply produce single images that one would be happy to hang on one's wall; they also use illustrative material to create animated views and whole movies in order to present the concepts and ideas of the best designers for our benefit.

The development of illustration from 1975 to 1994

Examples: 935 / 936 (1976)
911 (1984)
968 (1992)

» The basis for the creation of these illustrations was always a photographic record of all the technical details. As many as 300 detail and overall views of the subject would be taken. These would be supplemented by sketches or technical information from the engineering drawings. On the basis of this information a perspective design drawing was produced on translucent paper on the drawing board. Only then did the coloring take place, using a brush and gouache or acrylic paint. Finally the larger areas were given smooth modeling with an airbrush. The overall time spent would be up to 250 hours.

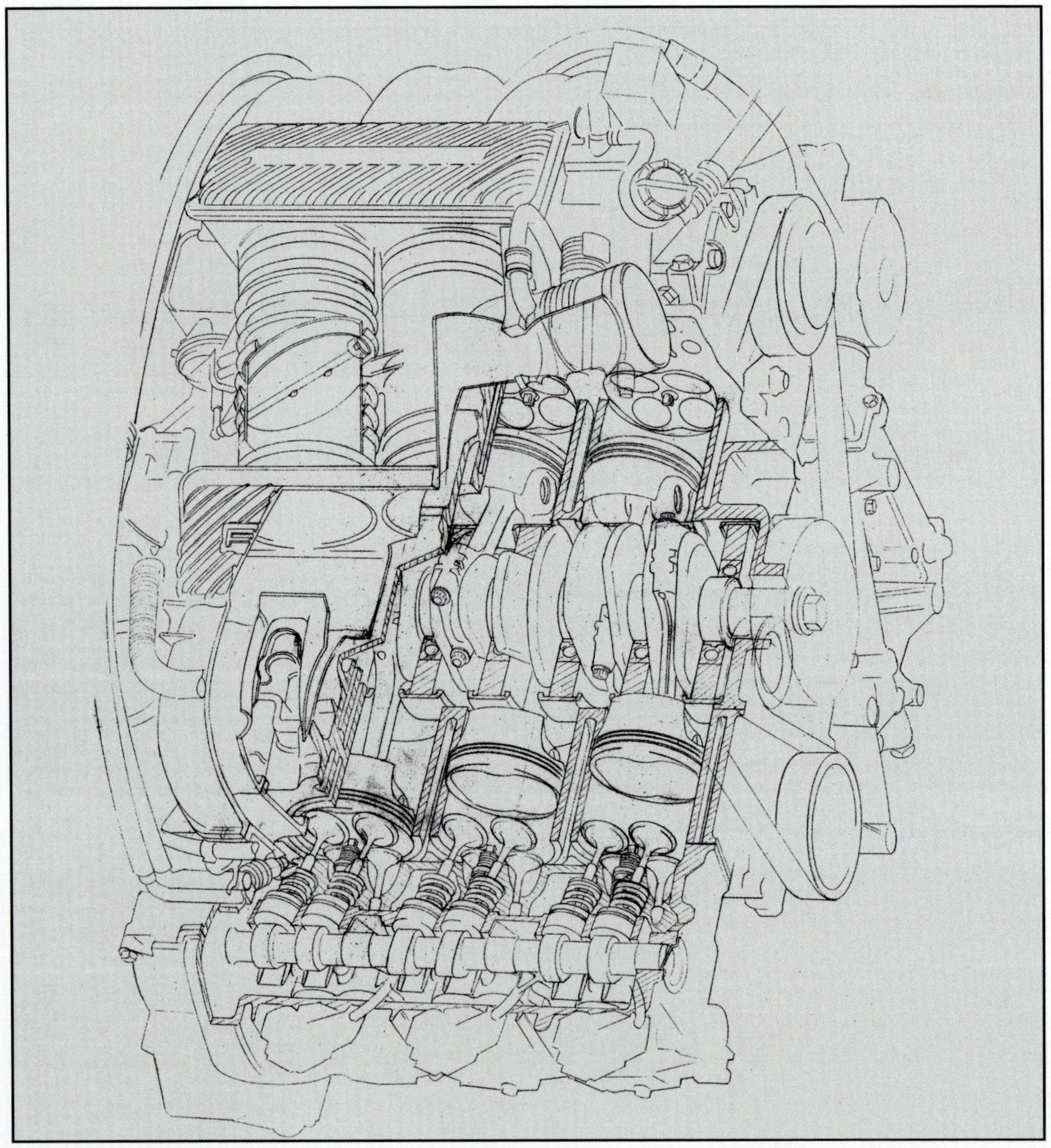

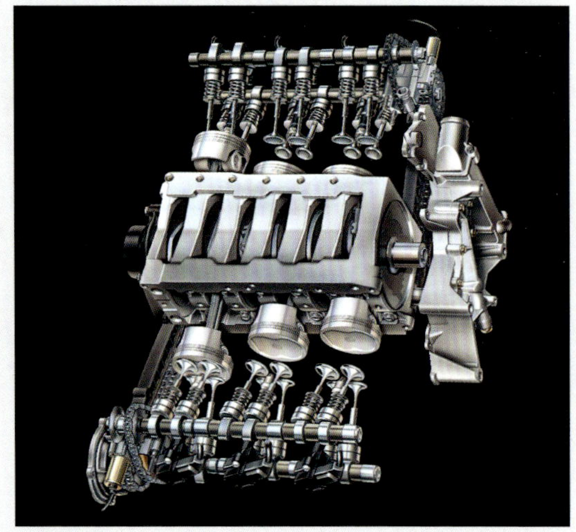

The development of illustration from 1995

Examples: 993 (1995)
996 (1997/98)

›› Here too the basis for creating the illustrations was the photographic record of all the technical details. The difference compared to earlier graphics (up to 1995) was the further digital processing carried out with a Quantel Paintbox, then the most powerful computer graphic system in the world. For the first time it was possible to create a picture of a complete vehicle by combining individual technical elements such as the chassis, wheels, engine, transmission, and body. From this, complete documentation could be created of everything from the chassis to the complete vehicle. In this way about 30 to 40 graphic images were created. The time taken for each project was about four to six months.

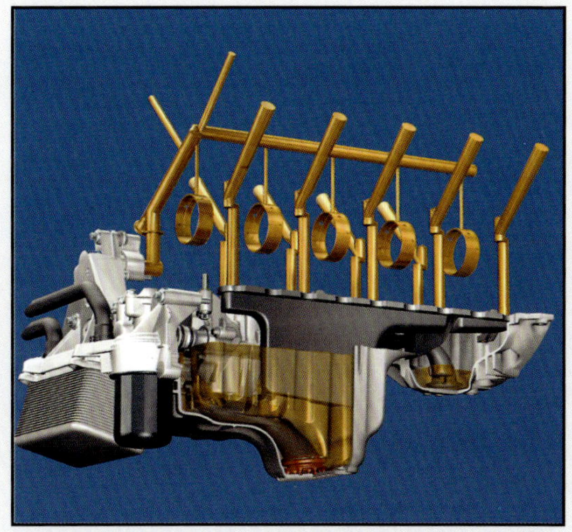

The development of illustration Cayenne 2001 / 2002

» The Cayenne was the first car that was completely created from 3D design data. This was possible because this vehicle was designed completely from scratch.

Here too a large amount of photographic information was needed in order to capture the visual appearance of the materials. The actual work then started with the 3D design data provided by Porsche, which were then converted by us into the appropriate data format for further processing. Using the photographic information collected, an extensive materializing of the data then took place. Properties of light and reflection had to be created for each component. After this the individual graphics were created by time-consuming computer rendering, and they were then combined in image compositions to produce the final result. In this way a vast image documentation was created. Because the creation of the vehicle involved technical modifications over a period of some time, this was the most comprehensive preparation for a complete vehicle that had taken place. Since then all other new vehicles have been created in 3D in this way. The date exchange process has been optimized and today materialization has already been applied to the data delivered.

MARTINI
PORSCHE
GOODYEAR
3
Shell
Shell
MARTINI
PORSCHE
DUNLOP
Shell
BOSCH
1

Racing cars

›› **Two fascinating models** from the world of racing, the Porsche 935 and 936, collected many world championship titles. While the 935 was based on the 911 Turbo and built with parts of the legendary 917 Dominator, making a name for itself in the endurance races of FIA-Group 5 (World Championship for Makes), the 936 was based on the racing technology of the models 908/3, and 917/10. It took the World Sportscar Championship for FIA-Group 6, winning every round.

The overall views of these two remarkable racing cars were drawn soon after their completion. They provide an amazing insight, and when comparing these two cars the design differences between them are readily apparent.

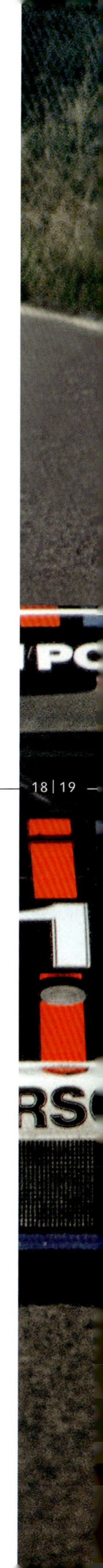

Porsche 935

A success with a flat nose

» **The legendary sports car** with the designation 911 provided the basis for the Porsche 935. To be precise, it was the 911 Turbo, also known as the 930, that was the basis for the cars entered in FIA-Group 5 of the World Championship for Makes in 1976.

In appearance it retained only the roof, and the windshield of the 911 range. The front of the car was a flat nose. The characteristic inward slope of the wings was dropped to improve the car's aerodynamics and the head-lights were moved into the front spoil-er. Enlarged wheel arches provided more room for wider wheels. The rear too was consid-erably modified com-pared to the production model. For one thing there was a large rear wing, prompted by a desire to improve aerodynamics that contributed to the car's maximum speed of up to 217 mph (350 km/h).

Porsche 935

» In comparison with the wide rear of the Turbo, the 935 had an even wider track. The rear wings were no longer rounded but angular in shape. There was an enormous air inlet in front of each rear wheel to keep the temperature down when in racing mode. Much more had been done inside the outer skin. The self-supporting steel body was fitted with fiberglass parts and as a result the weight of the car when empty was only 2,138 lb (970 kg). This powerful car was fitted with internally ventilated disc brakes, borrowed from the Porsche 917.

Compared to the production Turbo model, the 935 has a slightly smaller engine capacity of 174.3 cu in (2,857 cc) instead of 182.6 cu in (2,993 cc). Even so, the air-cooled six-cylinder boxer engine developed more than 260 bhp at 5,500 rpm. The 935, on the other hand, rotated at 7,900 rpm and had a maximum power output of 590 bhp, demonstrating the finest Porsche racing technology that was successful from the very beginning.

In 1976 the Porsche 935 dominated all the races in Group 5 of the World Championship for Makes. In 1977 the 935/77 version the power increased by 60 bhp to 650 bhp. In addition, the 935/77 Baby variation was entered in the lower division of the German Racing Championship. Its engine only had a capacity of 87.0 cu in (1,425 cc), driving this version that only weighed 1,653 lb (750 kg) to a top speed of 168 mph (270 km/h).

935

MARTINI RAC
MARTINI
MARTINI PORSCHE
MARTINI
MARTINI PORSCHE
DUNLOP
TIRES
DUNLOP
technical art

The long list of magnificent races run by the Porsche 935 was crowned by victories in the 24 Hours of Le Mans, the 24 Hours of Daytona, the 1,000 Kilometers of the Nürburgring, and the 12 Hours of Sebring. The works drivers included the future Formula 1 drivers Jacky Ickx, Jochen Mass, Rolf Stommelen, and Manfred Winkelhock, as well as Klaus Ludwig.

Also worthy of mention were the victories of Bob Wollek but he was never a works driver for Porsche. The further developments of the 935 by Kremer Motorsport were particularly successful. The victory of Porsche-Kremer in the great classic the 24 Hours of Le Mans in 1979 with drivers Klaus Ludwig, Bill Whittington, and Don Whittington was certainly one of the highlights of the car's career.

But Joest Racing and JLP Racing were also successful with the 935. From 1978 onward, only private teams were still entering the 935 in competitions, apart from Le Mans. There the 935 that competed was the 935/78 "Moby Dick" version. The engine was new: with 195.9 cu in (3,211 cc) it had a larger capacity, four overhead camshafts, and water-cooled cylinder-heads, developing up to 845 bhp at its maximum of 8,200 rpm. The aerodynamic overhang at the rear was 8 in (20 cm) longer while the wheelbase was 0.3 in (8 mm) longer. Weighing 2,260 lb (1,025 kg), the car had a top speed of 230 mph (370 km/h). This speed was reached on the Hunaudières straight in Le Mans.

Private customers continued to compete successfully with this model until 1981. A change in the regulations ended its career but it soon resurfaced in historic events for period cars.

The Porsche 935 in action: in 1976 its rivals had nothing to laugh about at the Nurbürgring

MARTINI PORSCHE
MARTINI
MARTINI PORSCHE
Shell
Shell
DUNLOP
Shell

Stoßdämpfer
MARTINI PORSCHE
MARTINI
MARTINI RACING
MARTINI PORSCHE
BILSTEIN
BOSCH
Shell

GOODYEAR
MARTINI PORSCHE
DUNLOP
MARTINI
Shell

Porsche 936

A Spyder that won Le Mans three times

» **In just a few months** in 1975 Porsche built a sports car to comply with the regulations of FIA Group 6 in the World Sportscar Championship. Benefiting from experience gained in developing the 908 and 917, and using of some parts from these models, Porsche was able to complete the project successfully. They built three cars in all. The basis was an aluminum tubular space frame concealed within a fiberglass body. Its low nose with its large air inlet greedily sniffed the asphalt while the long rear supported a large wing. Behind the driver, who sat very far forward, was a large engine air box projecting upward, ensuring a good supply of air for the turbocharger intercooler. The curb weight of the racing car ready for driving was a mere 1,543 lb (700 kg). With a wheelbase of 94.5 in (2.400 m), the car had an overall length of 167.3 in (4.250 m). The car was just under 78.7 in (2 m) wide, and only a little more than 39.4 in (1 m) high. The front track was 62.2 in (1.580 m), 2.76 in (70 mm) wider than the rear track.

Porsche 936 Spyder

›› The rear housed a large 130.7 cu in (2,142 cc) air-cooled boxer engine with turbocharging. The turbo equivalency factor of 1.4 meant that the car conformed to the class regulations for cars of up to 3 liters in capacity.

The extremely short stroke of this power unit – bore x stroke 3.23 x 2.60 in (83 x 66 mm) – was combined with a very low compression ratio of 6.5:1. In 1976 the engine developed 520 bhp at 8,000 rpm. Two chain-driven overhead camshafts looked after the inlet and exhaust cycle. Unlike the Porsche 935, drivers of the 936 had five gears available instead of four, and the power transmission was completed by a self-locking differential. The top speed was 205 mph (330 km/h). Internally ventilated steel disc brakes slowed the racing car. At the front it had transverse control arms with push rods and a stabilizer bar, and at the rear lower A-arms each with upper and lower longitudinal push rods. Variable rate coil springs and telescopic shock absorbers ensured that the 15-inch tires are always in perfect contact with the road surface. In the 1977 season the 936/77 Spyder version of the car had an extra 20 bhp. Two of them Porsche converted by fitting smaller turbochargers with better response. There were also a few smaller modifications to the chassis. The wheelbase was increased by 0.4 in (10 mm) and the body was 1.97 in (50 mm) shorter. At the same time the front and rear tracks were narrower by 1.97 and 1.18 in (50 and 30 mm) respectively.

936 Spyder

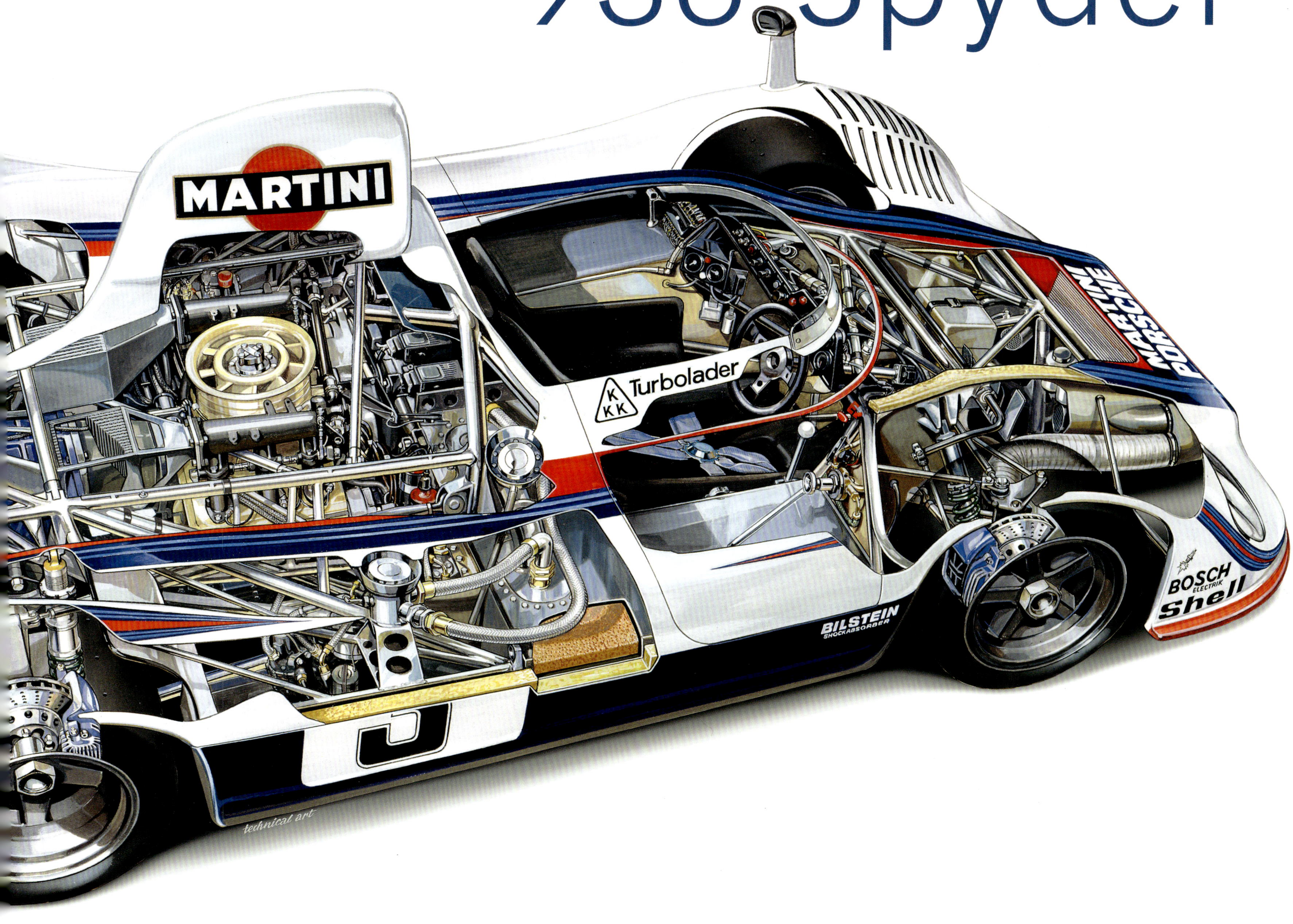
MARTINI
Turbolader
KKK
MARTINI
PORSCHE
BOSCH
ELECTRIK
Shell
BILSTEIN
SHOCKABSORBER
technical art

The Porsche 936 dominated the World Sport Car Championship from the very start. The 1976 title went unchallenged to the maker from Zuffenhausen. In 1976 Porsche also won the 24 Hours of Le Mans with the car with chassis number 002. In 1977 it was the Porsche with chassis number 001 that crossed the finishing line of this classic endurance race in first place. In both years the Belgian driver Jacky Ickx was behind the wheel of the winning car, in 1976 accompanied by the Dutchman Gijs van Lennep and in 1977 by Jürgen Barth and Hurley Haywood. In 1978 the 936 just missed victory when Porsche "only" finished in second and third places, but 1979 was even worse in that the two Porsches entered in the race had to retire. But the career of the 936 was not yet over. Admittedly Porsche had added the car to its museum, but one car, later identified as a "genuine" replacement chassis with number 004, entered by the Joest Racing team in 1980 and driven by Reinhold Joest, and Jacky Ickx, finished second. In 1981 the 936/81 Spyder won yet another victory in Le Mans, again with the Belgian driver Jacky Ickx at the wheel. But the engine had been modified, and now had a capacity of 161.7 in (2,650 cc). With four overhead camshafts driven by spur gears and water-cooled cylinder heads, the car had a maximum power of 620 bhp at 8,000 rpm. Weighing 1,653 lb (750 kg) and just under 187 in (5 m) long, the car had a maximum speed of 224 mph (360 km/h). But the drivers still had to go without something: there were only four forward speeds, and no differential lock for the powered rear wheels.

A successful team: Martini Racing, and Porsche with (from right): Jacky Ickx, Jochen Mass, Manfred Janke, Rolk Stommelen, and Manfred Schurti; back, second from right, Jörn Pugmeister.

PORSCHE
STUTTGART
MARTINI
RACING
Turbolader
BILSTEIN
BOSCH
WIEDERHOLD
DUNLOP
Shell
MARTINI PORSCHE
MARTINI RACING
MARTINI
DUNLOP
Shell
BOSCH
MARTINI PORSCHE
GOODYEAR
6
MARTINI
PORSCHE

S-PR 986
S-PR 191
S-UV 333
S-PR 981

Classics

» **Because the first** cutaway drawings of Porsche models were created in the mid-1970s, no such drawings of earlier vehicles exist. The drawings of Porsches made in association with Volkswagen – the 924 and its variants – are becoming increasingly esteemed by enthusiasts. Its successor, the 944, has also slowly outgrown its status as a mere secondhand car. The more recent four-cylinder with the designation 968 needs a little more time to do that but the clock is still ticking away inexorably.

Another front-engine Porsche forms a bridge between the classic "Youngtimer" (a term used for secondhand cars of the 1980s), and secondhand cars: the 928 with its eight cylinder power unit is shown in its various evolution stages, before it was raised to the status of a modern classic.

Admittedly the Carrera GT is still very young but this sports car is fascinating with its ten cylinders, and sophisticated technology. After its status as a car sought-after by enthusiasts, it is now on the way to becoming a classic. This is also the case with all Porsches with the designation 911 since 1964. This model perfectly illustrates the technical developments and the changes in Technical Art's graphic conversion techniques, from the G-model in 1975 to the 997 series, replaced in 2011 by the current 991 series.

Porsche 924

An ever-increasing appreciation

» **After almost 120,000** had been built, the production of the VW-Porsche 914 came to an end in 1975. In 1970 Porsche developed a successor for this sports car, using many parts from Volkswagen and Audi models. This collaboration did not last because the successor was too expensive for Volkswagen's range. So Porsche arranged the manufacture of the 924 itself, subcontracting to Audi in Neckarsulm. Compared to the classic Porsche models, and also the VW-Porsche 914, the difference was considerable.

It had an inline engine (from the Audi 100) with water-cooling fitted at the front, instead of an air-cooled boxer engine at the rear. In the eyes of the clientele, this was no longer consistent with the concept of a true Porsche. The further use of Volkswagen and Audi parts fueled this impression.

Increasing appreciation

›› In spite of this initial reserve on the part of the customers, the 924 was a success for Porsche. Production, and delivery started in February 1976. Priced at just over 23,000 DM, the 924 was more than 12,000 DM less expensive than the legendary 911 but it was neither perceptibly less powerful, nor slower. The transaxle technique with the gearbox integral with the rear axle together with the front-mounted engine behind the front axle ensured a well-balanced weight distribution and excellent performance on the road. With its large rear hatchback, the 2+2 seater 924 also proved practical.

With a four speed-gearbox and narrow 165 R 14 tires, the car had a top speed of 124 mph (200 km/h). By the fall of 1976 Porsche was already upgrading the model and increasing the level of equipment A five-speed gearbox and standard tires were introduced in 1977 and 1978 respectively. In 1979, the 924 was fitted with a turbo engine. Its 170 bhp, later raised to 177 bhp, made the car even livelier. Additional cooling air slits were fitted at the front between the elegant retractable headlights as well as bigger tires, now 185/70 R 15, drawing attention to the model's increased power from the outside. The rear drum brakes were replaced by discs.

The next, more sporty model, the Porsche 924 Carrera GT, produced in a limited edition of 400, followed in 1980. With a turbocharger with intercooler, the Audi engine now developed 210 bhp. The car could reach 62 mph (100 km/h) in seven seconds from a standing start and had a top speed of 149 mph (240 km/h). The model had 215/60 R 15 tires and visibly wider wings, increased by 2.4 in (6 cm) at the front and 3.1 in (8 cm) at the rear. By April 1981 Porsche had built a total of 50 of the 924 competition models, the Carrera GTS (245 bhp) and the Carrera GTR (280 bhp). The 924 Turbo was produced until 1982 but it was no pleasure for drivers because of problems with the turbocharger. The normal model with an Audi fuel-injection engine oft 125 bhp remained in production until 1985. The US and Japanese versions had to make do with 112 bhp.

924

This advertisement highlights the "gearbox in the back, engine in front; transaxle in the middle; greater driving safety: active and passive."

924 S

Now with a Porsche engine

The car known as the Porsche 924 S, launched in 1985, had a completely different power unit. It was a four-cylinder inline engine with a capacity of 151.3 cu in (2,479 cc), built by Porsche. Compared to the engine in the parallel Porsche 944, its power output was reduced from 163 to 150 bhp. But by 1987 the 924 S and the 944 had the same power output while the 924 boasted a very slightly higher top speed – 2.5 mph (4 km/h) more, giving it a maximum speed of 138 mph (222 km/h) – on account of its narrower body without flared wheel arches. The 944 had a wider track at the rear and a 21-gallon (80-liter) fuel tank, an increase of nearly 4.8 gallons (18 liters). Production of the Porsche 924 ended in 1988. The last version of the sports car was the 924 S Exclusive, priced at 52,950 DM, produced in a limited edition of just under 1,000, known by the words "Le Mans" adorning its sides. These cars are particularly favored by connoisseurs.

The earlier special models of the 924 are also much sought after by collectors. In 1977, in recognition of Porsche's sponsor, an edition of "Martini" cars was produce. About 1,000 of these cars with white-painted steel rims were built for Europe and 2,000 for the USA. The 924 Sebring, produced in a limited edition of 1,400, was only sold in America (1980). In the same year Porsche produced 1,030 "Le Mans" special models, painted white with a yellow, black, and red decorative band. In 1981 the 924 Ferry Porsche and 924 Weissach limited editions were launched.

Porsche 944

The fastest four-cylinder cabriolet in the world

» **The Porsche 944** celebrated its world premier at the Frankfurt Motor Show. Production began in January 1982 as planned. Visually the model had a strong similarity with the 924 Carrera GT with its conspicuous front wings. The corresponding widening at the rear no longer looked like an afterthought but flowed as part of a harmonious whole. On the whole the car was more impressive than the narrower 924. Nothing had changed in the technical concept of the engine at the front behind the front axle and the gearbox integral with the driving axle, the so-called transaxle configuration.

With half a V8 engine

» The new engine appears to have played an important part in the success of the model. Porsche had divided the V8 of the 928 in two and developed a four-cylinder inline engine with a capacity of 151.3 cu in (2,479 cc), developing 163 bhp at 5,800 rpm. Two balance shafts reduced vibration. It could be proud of its maximum speed of over 143 mph (230 km/h) and its fuel consumption was pleasantly low. The maximum torque was 151 ft lb (205 Nm) at 3,000 rpm. The standard five-speed gearbox could be replaced by a three-way automatic gearbox on special order, and for an extra charge.

With a cost price of 38,900 DM, the 944 was a good 5,300 DM less expensive than the equally powerful 924 Turbo, the production of which had come to an end in July 1982. Porsche gave a seven-year guarantee against rust corrosion of the galvanized all-steel body. There was plenty of room for luggage under the large rear window of the hatchback, especially when the rear seats were folded back. Rather unusual for a sports car was the possibility of fitting a trailer coupling: up to 2,645 lb (1,200 kg) of weight could be towed. In addition, it was even possible to order a roof rack, specially designed for the 944.

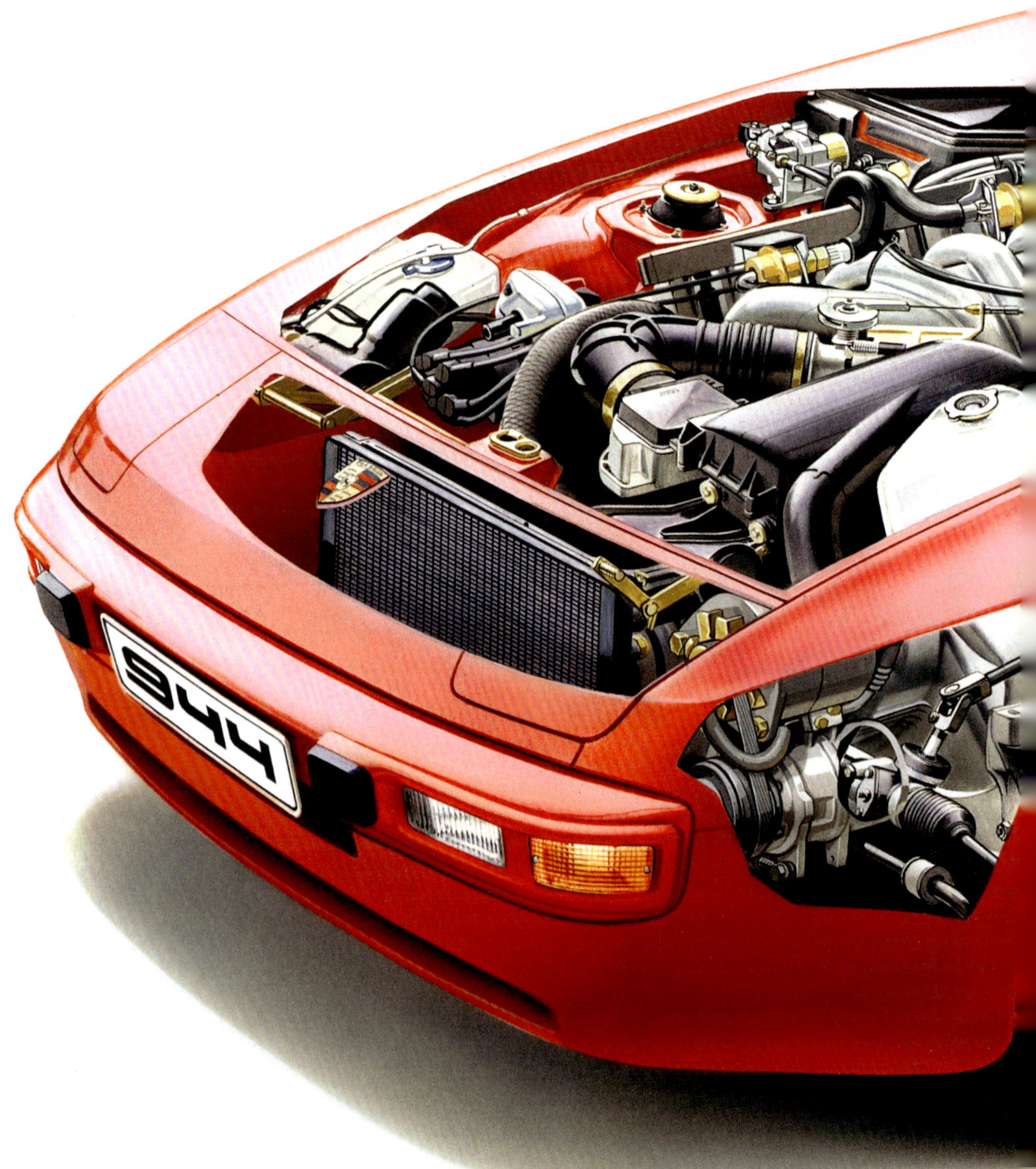

944

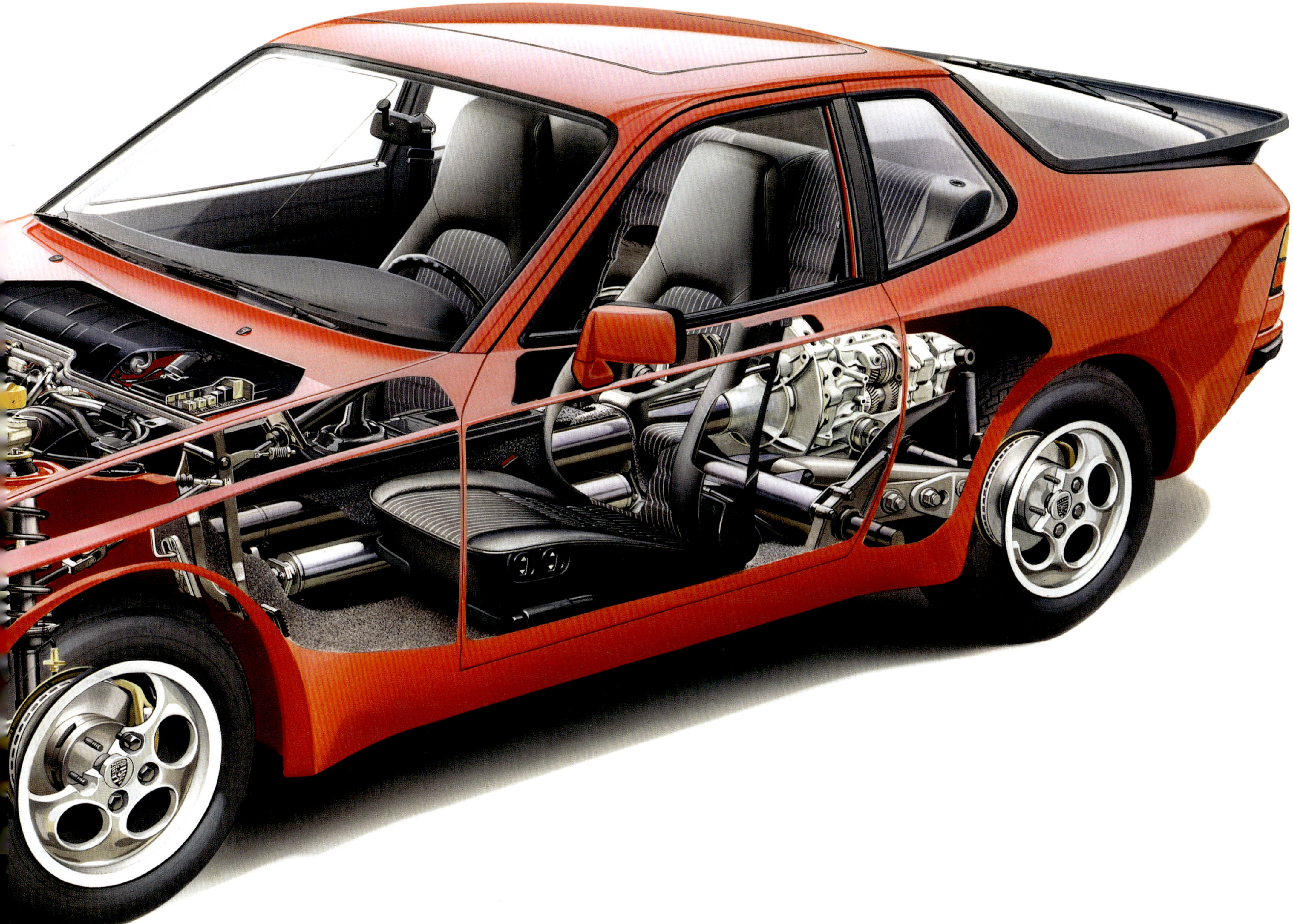

944 S

A real Porsche, built by Audi

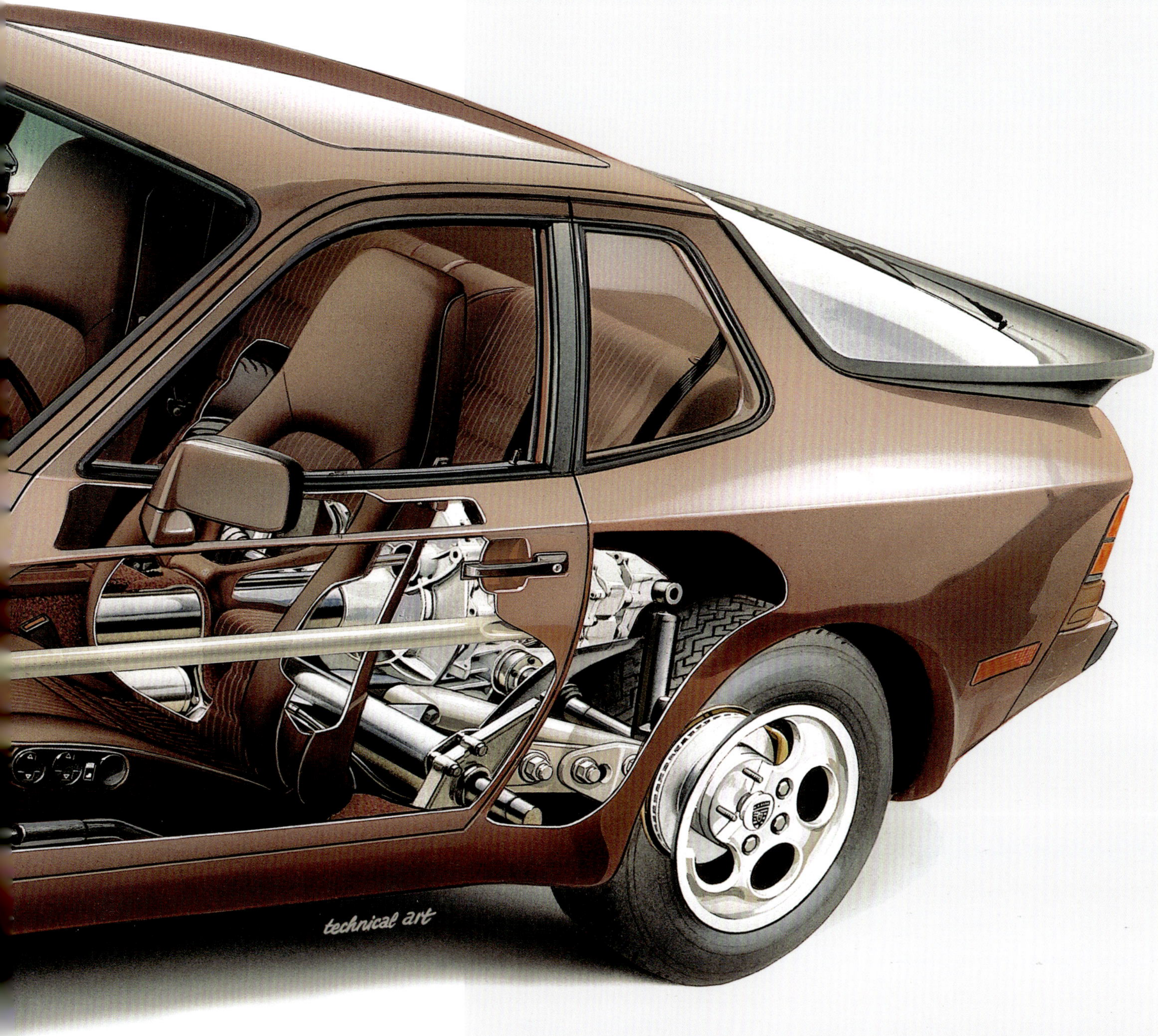

» Porsche ensured ongoing sales success by systematic upgrading. In 1985 the car was fitted with an exhaust gas catalytic converter but this reduced the power to 150 bhp. A flush windshield, glued in place, with am integrated aerial, a newly designed instrument panel in the style of the Porsche 928, a fuel tank with capacity increased from 16 to 21 gallons (62 to 80 liters) as well as new aluminum wheels added to model's attraction. The Turbo 944 was launched in July 1985. Priced at 74,000 DM, the 220 bhp model was almost 22,000 DM more expensive than the normal 944, and 32,000 DM more than the parallel 924 S. The power of the Turbo was delivered through forced induction by a KKK 26 turbocharger with intercooler. But it still had overhead camshafts driven by toothed belts. In the 1988 version the power output was 250 bhp. In 1986 an anti-lock braking system (ABS) became available for an extra charge, only becoming part standard equipment in 1989. In 1986 Porsche developed the 944 S with the Bosch L-Jetronic fuel injection system instead of the Bosch Motronic. Four valves per cylinder were operated by twin overhead camshafts driven by toothed belts like twin roller chains.

Built as a cabriolet since 1989

» Until 1988 the power output of the 944 S was 190 bhp at 6,000 rpm, then between 1988 and 1991 it rose to 211 bhp in the 944 S2, and at the same time it had a higher torque of 296 ft lb (280 Nm) as compared to 170 ft lb (230 Nm) at a lower rotational speed (4,100 rpm instead of 4,300 rpm). This was because the capacity had increased to 182.5 cu in (2,990) cc as a result of a wider bore and longer stroke.

The 944 was also available as Targa. Its sunroof could be moved backward or completely removed and stored in the luggage compartment.

Finally, in January 1989, Porsche built a cabriolet version of the 944 that had been launched in 1985. The conversion and final assembly were entrusted to the American Sunroof Company (ASC) in Weinsberg, just a few miles the north of the parent plant in Zuffenhausen. Initially only the 944 S was available in the cabriolet version. Just under 7,000 were sold until its production came to an end.

944 S2

944 Turbo

Small engine, great power

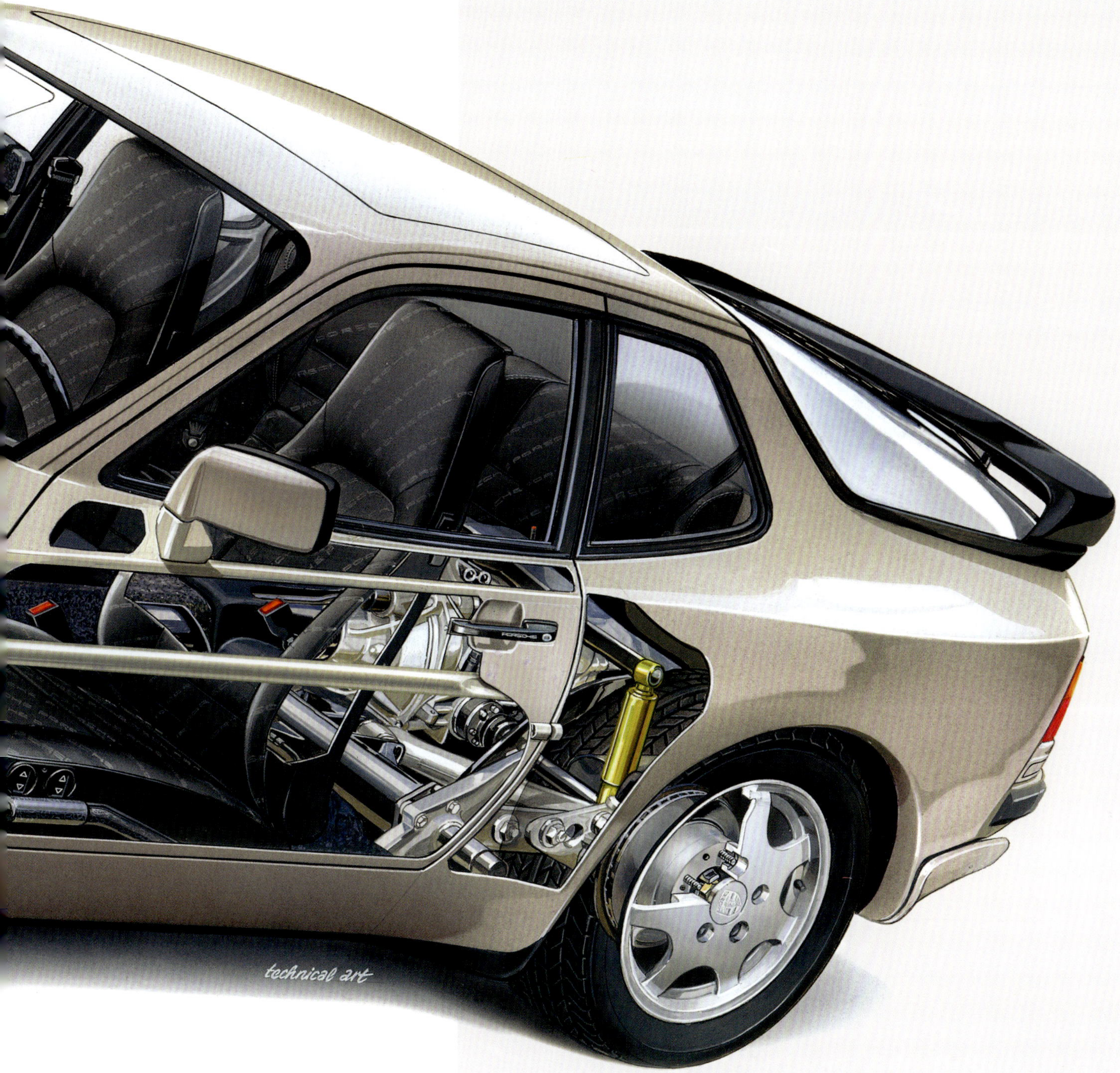

›› In 1990/91 Porsche also produced a special edition of 528 examples of the 944 Turbo cabriolet. This cost 11,000 DM more than the coupé; until 1990 an electric soft top actuation had been available for an extra cost. The American Sunroof Company (SC) welded a second floor to the body as reinforcement. The windshield frame was shortened by 2.4 in (6 cm) and the large window at the rear was replaced by a normal trunk lid. Also worth mentioning are the previous history and racing career of the 944 Turbo. Porsche had already tested the technology when the car took part in the 24 Hours of Le Mans in 1981. Designated the 924 GTP, the car was driven by Walter Röhrl/Jürgen Barth who finished seventh overall. Three years later Porsche displayed the full speed strength of the exhaust gas catalytic converter in a 24-hour race in Ohio. A year after the start of production of the 944 Turbo, Porsche launched the 944 Turbo Cup. These specially modified vehicles had different suspension components and of course larger tires, as well a built-in safety-cage. In 1987 it was fitted with ABS.

Porsche 968

The crowning of the four-cylinder front-engine series

» **The successor** to the Porsche 944 cars manufactured between 1981 and 1991 had a completely new numeric designation, even though its close relationship with the 944 is very obvious. The model was given the designation 968, and between 1991 and 1995 some 11,245 were built. It was not a great success although it was replacing the proven and tested 944 S2.

But the 968 could not dissociate itself completely from the poor image of Volkswagen-Porsche. Sales figures were disappointing: besides the 5,731 coupés and 3,959 cabriolets, there were also 1,538 examples of the 968 CS that did not have the rear bench seat, electric window regulator, airbags, or rear window wiper; it was 110 lb (50 kg) lighter than the coupé.

The visual appearance was more sophisticated than that of its predecessor. The rear was reminiscent of the eight-cylinder 928 while the front borrowed elements from the 911 type 993 but had open retractable headlights.

For the first time with VarioCam

» Under the body, the chassis still had MacPherson strut suspension at the front with the lower transverse control arm, coil springs, and stabilizers. The semi-trailing arm rear suspension had a diagonal torsion bar, hydraulic shock-absorbers, and stabilizers. The 4-valve 3-liter engine of the 944 S2 had a slightly higher compression ratio (from 10.9 to 11:1) as well as the first VarioCam, a hydraulic camshaft adjusting device, and a newly optimized induction system giving 240 bhp at 6,200 rpm. The torque rose from 206 to 224 ft lb (280 to 305 Nm). The declared maximum speed for the coupé and cabriolet was 157 mph (252 km/h).

Both the wheelbase wheel track remained unchanged from the 944. The diameter of the internally ventilated disc brakes was the same as that of the S2, and Turbo versions. The six-speed gearbox was new while a four-speed Tiptronic automatic transmission was available as an option.

968

A self-locking differential was also available at additional cost. With its starting price of just under 90,000 DM, the 968 still cost 5,000 DM more than its predecessor. The model, available as a cabriolet from the beginning, remained beneath the 100,000 DM threshold. On the other hand, the slimmer Club Sport model, postponed to 1993, was available for 77,500 DM. The 17-inch tires were 225/45 at the front and 255/40 at the rear, and the brake discs were slightly larger.

The sports highlights of the 968 series were the Turbo S and Turbo RS models, but since only fourteen and four respectively were built, they were not marketed. The Turbo S had a cylinder head with a single overhead camshaft and eight valves. But a turbocharger with inter-cooler gave it a maximum power output of 305 bhp at 5,400 rpm, with the low compression ratio of 8.0:1. Its torque peaked at 369 ft lb (500 Nm) at 3,000 rpm. The 968 Turbo S had 18-inch wheels with tires of 235/40 at the front and 265/35 at the rear. Accelerating from 0 to 124 mph (0 to 200 km/h) in just ten seconds, it was very much faster than the normal 968. Its maximum speed was 174 mph (280 km/h).

Like its predecessor, the 968 impressed with its high quality. It was economical, fast, and reliable: even driving more than 185,000 miles (300,000 km) would pose no problem with proper care and use. Not least because of the excellent precautions against rust represented by its galva-nized body, the four-cylinder front-engined Porsche kept its value. The transition from secondhand car to being a sought-after, popular young classic has nearly been completed. Besides the cabriolet version of the 968, the very spartan CS model is also highly regarded. Exactly 3,959 cabriolets were built, of which 1,538 were the CS.

» The three photographs above were taken during the preparation of the cutaway drawings by Technical Art. Good lighting of the objects and correct perspective were essential. The studio floor did not interfere as a background. Photographs were taken of the axles, drive train, and engine in position; the pictures then served as masters for the drawings made by hand. The drawings of the 968 were last of Technical Art's Porsche illustrations to be drawn by hand without electronic assistance. A comprehensive description of the three different ways of making illustrations can be read on page 10.

Porsche 928

The V8-GT with transaxle

» **The large new sports car** from Porsche never inherited the iconic status of the 911. In the early 1970s Porsche started work on the development of the new model that was presented at the 1977 Geneva Motor Show with the designation 928.

The new top model of the marque was designed in the same way as the smaller 924: the engine at the front and the gearbox in the transaxle position, that is, integral with the driven rear axle. The modern design of the four-seater coupé 177 in (4.5 m) long had open retractable headlights at the front and a glazed rear window of generous size.

Doors, engine hood, and front wings of aluminum kept the weight of this luxury car with electrically operated seats, mirrors, and windows down to 3,197 lb (1,450 kg). As a result the newly developed, liquid-cooled aluminum V8 engine had it easy: 62 mph (100 km/h) could be reached from a standing start in less than seven seconds while the car had a maximum speed of 143 mph (230 km/h). In this complex chassis with independent suspension all round, the rear axle was particularly interesting because it was built in such a way that during normal driving it automatically applied toe-in to compensate for oversteer.

The V8-GT as a 911 replacement

» The construction of the Porsche 928 fast touring car was completely different from that of the rear-engined 911 sports coupé. Mounted over the front axle, the 90° V8 engine had a capacity of 273 cu in (4,474 cc) and developed 240 bhp. Each cylinder bank had an overhead camshaft operating two valves per combustion chamber and driven by timing belts. The maximum torque of 258 ft lb (350 Nm) was reached at a moderate 3,600 rpm. The power was conveyed to the rear wheels via the five-speed gearbox. There was also a three-speed automatic transmission available as an option. The car had four internally ventilated disc brakes with floating calipers.

The driving behavior could not be criticized as the notorious tail slide of the 911 had been, because the "Weissach rear axle" with lower transverse control arms and longitudinal push rods optimized active driving safety. The transaxle construction ensured a well-balanced weight distribution. Anti-dive and anti-squat systems compensated for pitching when braking and sagging when accelerating. As a result the 928 was less a sports car and more a comfortable Gran Turismo. But in the end the 928 did not turn out to be a replacement for the 911. Admittedly it was accepted as a true Porsche but sales did not meet expectations.

928

Powerful and sophisticated

» In 1979 the basic model of the Porsche 928 was joined by a more powerful version, the 928 S. It was presented at the Frankfurt Motor Show. The bore was increased by 0.078 in (2 mm), bringing the capacity to 284.6 cu in (4,664 cc). Together with the increase in compression ratio from 8.5 to 10.0:1, this improved the power output to 300 bhp at 5,900 rpm. The time from 0 to 62 mph (0 to 100 km/h) was two-tenths of a second faster and the top speed was 250 km/h. The price premium over the basic model did not prove to be a real obstacle to sales.
The 928 S also had several enhanced external features now that there was more power under the hood. The right-hand mirror was standard, as were the redesigned alloy 7J x 16 wheels. The aerodynamics were completed by a small spoiler and a rear-mounted spoiler lip, made of rubber; together these increase the downforce. In 1980 the cost of the 928 S was over 75,000 DM. In 1982 the final series of the normal 928 cost almost 65,000 DM and by 1985 the price of the S variant has climbed to 109,000 DM.
In 1986 the last 928 S, fitted with a 305-cu in (5-liter) engine, was dropped from the program. Presented in 1987, the 928 S4 with an M 28/41 engine with four-valve technology and twin overhead camshafts per cylinder bank took over from it.

928 S

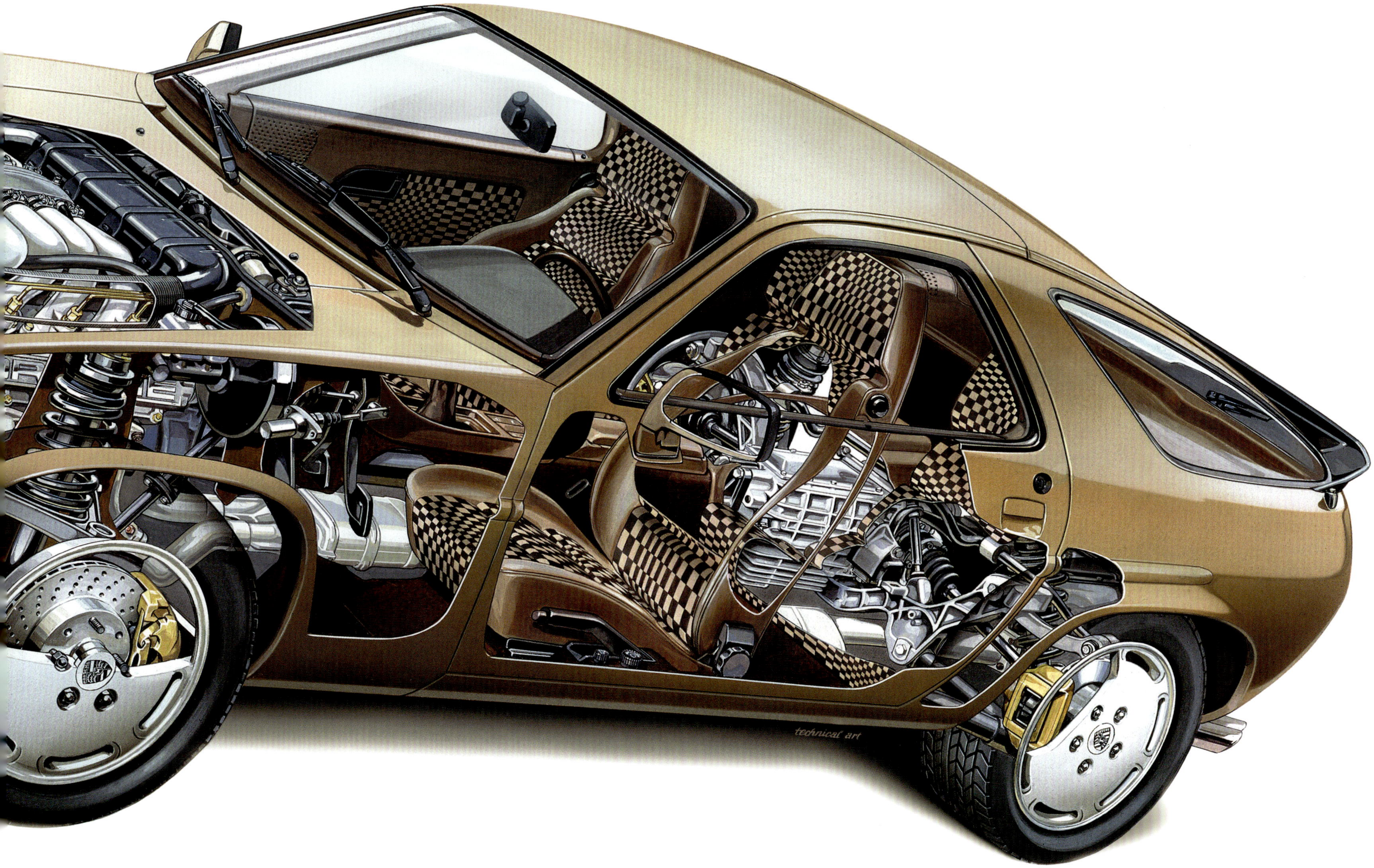

Modified lights

» In 1985 the appearance of the front-engined Porsche was slightly altered. The front became more rounded, the fog lights were modified, and additional headlights were fitted. New air vents were added in the front spoiler as cooling air inlets for the brakes. New light clusters and a larger spoiler adorned the rear. With the new underbody cladding and the newly developed control of cooling air, the car now had a drag coefficient of 0.34. Motor-operated flaps controlled the cooling air so that only the amount of air needed was allowed to pass through.

The 928 S4 again had a bored-out engine with a capacity of 4,302.5 cu in (957 cc), developing 320 bhp at 6,000 rpm. At the same time the torque has risen to 317 ft lb (430 Nm) at 3,000 rpm. The number of camshafts was doubled to correspond with the number of valves. With a maximum speed of 168 mph (270 km/h), this model had new brakes with 4-piston fixed calipers. Usually clients choose the 4-speed automatic gearbox instead of the 5-speed manual gearbox. But automatic transmission was not available for the 928 S4 Club Sport that was supplied exclusively with a manual gearbox, and with more spartan fittings, so that it was about 264 lb (120 kg) lighter. This further improved the acceleration and led to a slightly higher maximum speed.

928 S4

Sportier with twin exhaust pipes

» The 928 GT was launched as a sportier sibling of the S4. The external differences were minimal. It no longer had protective moldings along the sides like the S4. The wheels at the front and rear were 1 in (2.5 cm) wider, and as a result had the same dimensions as the Club Sport model. Besides a clear GT signature, the exhaust system with twin exhaust outlets very much reflected the increased power. But unlike the Clubs Sport 928, the model was not spartan in its fittings because it had every luxury accessory one could think of, apart from an automatic gearbox. The power output, now increased to 330 bhp, could only be transmitted through the five-speed manual gearbox. The model had a maximum speed of 171 mph (275 km/h).

Further updating of the series took place in the fall of 1989 when the S4 model was also fitted with the twin exhaust pipe system. It was also fitted with the Porsche self-locking differential (PSD) with variable ratio limited slip from 0 to 100%. As a result the driving performance was even more stable and very docile even at the limit of stability; it reduced the wear and tear when maneuvering while also improving the traction.

928 GT

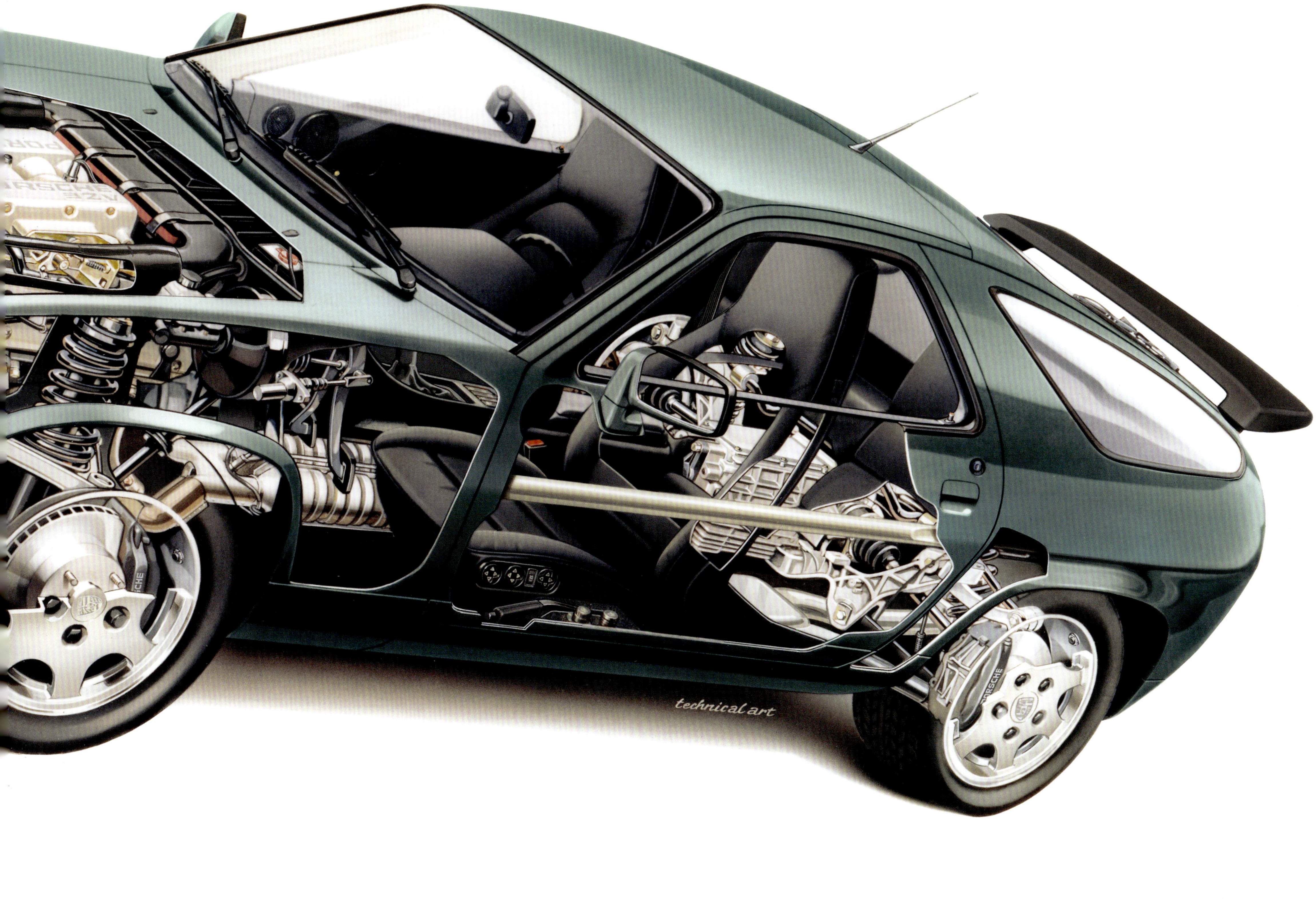

The end after 18 years

» After some 58,225 examples of the 928, 928 S, 928 S4, Club Sport, and 928 GT had been built, Porsche launched the 928 GTS to replace all the previous models. Porsche again increased the engine capacity of the new model, launched in the fall of 1991.

In this case the bore remained at 3.94 in (100 mm) but the stroke increased from 3.11 to 3.38 in (78.9 to 85.9 mm), giving a capacity of 5,329.3 cu in (397 cc). The maximum power output was 350 bhp at 5,700 rpm and, compared to the 928 GT, the maximum torque rose from 317 to 369 ft lb (430 to 500 Nm) at a slightly higher 4,250 rpm.

Visually what was most striking was the increase in the tire size to 17 inches (previously 16 inches) as well as the uninterrupted band of lights at the rear. The rear view mirrors were also newly designed. The car was 2.1 in (5.4 cm) wider than the GT and the heaviest of all the 928s with a curb weight of 3,574 lb (1,620 kg). It remained almost completely unchanged from 1991 until its production came to an end in 1995, by which time it cost some 156,050 DM.

928 GTS

Porsche Carrera GT

An uncompromising symbol of technology

» **The Carrera GT** was a fascinating super-sports car. The engine, chassis, and other technical parts made it suitable for a 24-hour endurance race.

But the Carrera GT was built by hand in Porsche's new Leipzig factory as a homologated vehicle in the fall of 2003, and was launched on the market at a price of 450,000 Euros. Of the 1,500 initially planned, only 1,282 had been built by 6 May 2006, including pre-series and close-to-production prototypes.

This emblematic symbol of tech-nology revealed a few features that had not previously been included in production cars.

Former World Champion rally-driver Walter Röhrl played an important part in its development. Test drives on the home circuit at Weissach and more especially on the northern loop (Nordschleife) of the Nürburgring were the in which chassis and suspension tuning were carried out.

As a result the car displayed racing-car road performance and handling characteristics. But clients were also offered a 5-piece set of luggage in leather, each piece fitted into its respective space, and matching the color of the leather-upholstered interior chosen by the client. The route navigation system and air-conditioning were built in on request at no extra cost while the color could also be chosen by the client at no extra cost.

Ten cylinders inside the hood

›› By choosing a monocoque chassis that had never been used in production cars before, Porsche was treading new ground. The chassis and drive assembly supports were made of carbon-fiber-reinforced plastic. The front and rear were designed to absorb impact in the event of a crash. Strong window frames and B-pillars ensured the necessary safety in the case of a rollover.

The V10 engine was mounted longitudinally in front of the rear axle. The short-stroke engine – bore and stroke 3.9 x 3.0 in (98 x 76 mm) had a capacity of 349.9 cu in (5,733 cc) and developed 612 bhp at 8,000 rpm. The engine had a very high compression ratio of 12.0:1. The maximum torque of 435 ft lb (590 Nm) was reached at 5,750 rpm, and the maximum rpm was 8,400 rpm. There were two chain-driven overhead camshafts per cylinder bank. They each operated four valves per cylinder by means of bucket tappets. The initial plan of developing it as a racing car was never realized. But at least in the fall of 2004 the car was able to set the lap record for homologated cars on the northern loop (Nordschleife) of the Nürburgring. The time of 7 minutes 32.44 seconds corresponded to an average speed of 164 km/h.

All the drawings of the Carrera GT shown here are based on data from 3D design drawings.

Carrera GT

Ceramic clutch and push rods

» One of the most important innovations on this Carrera GT – which had a maximum speed of 208 mph (334 km/h) and could accelerate to 124 mph (200 km/h) in 9.9 seconds – was its ceramic twin-plate clutch. Its official name is the Porsche Ceramic Composite Clutch (PCCC). In spite of being extremely compact, just 6.65 in (169 mm) in diameter, this could cope with the enormous forces. Also new in the transmission area were the Flexband gear change cables of the transverse six-speed gearbox. Instead of the usual steel cable, Porsche here used a stainless steel one that ensured high precision during its whole life cycle.

The suspension of the Carrera GT was also very interesting. The double wishbones, both at the front and rear, were operated by push rods. The front shock-absorbers rested on the carbon-fiber-reinforced plastic sub-frame. The corresponding rear shock-absorbers were mounted longitudinally. The standard equipment also included the internally ventilated ceramic brake discs with the front brake calipers enclosing the discs from behind and the rear brake calipers enclosing the discs from the front. These enabled the vehicle to come to a halt – thanks also to the ABS and brake power limiter – from 124 mph (200 km/h) in just over four seconds.

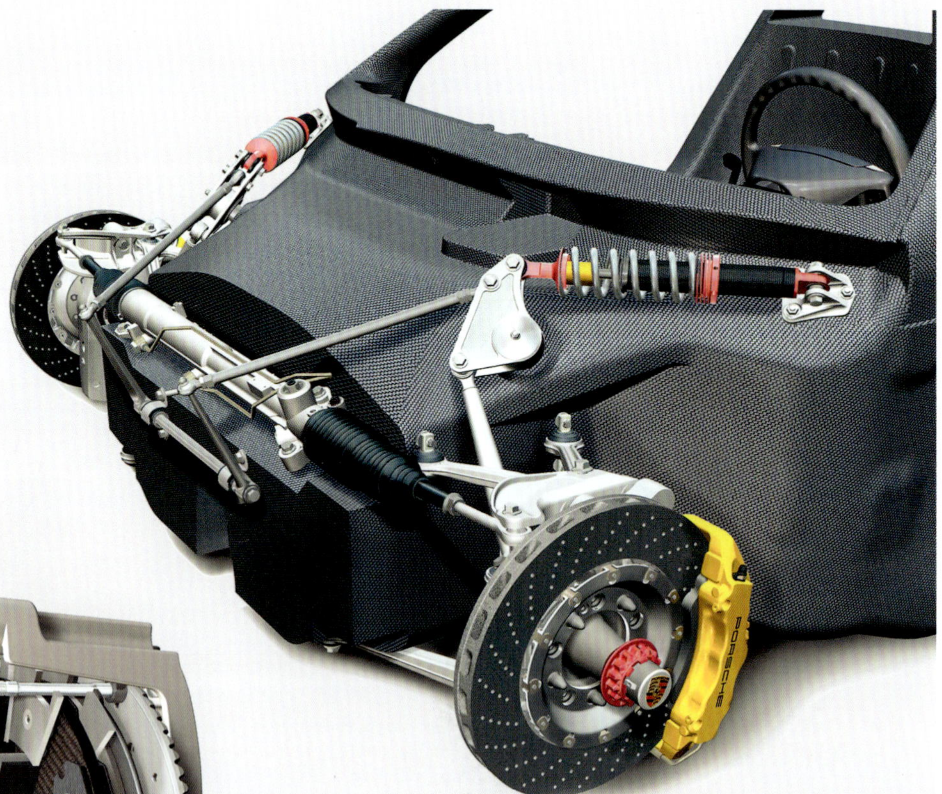

Carrera GT

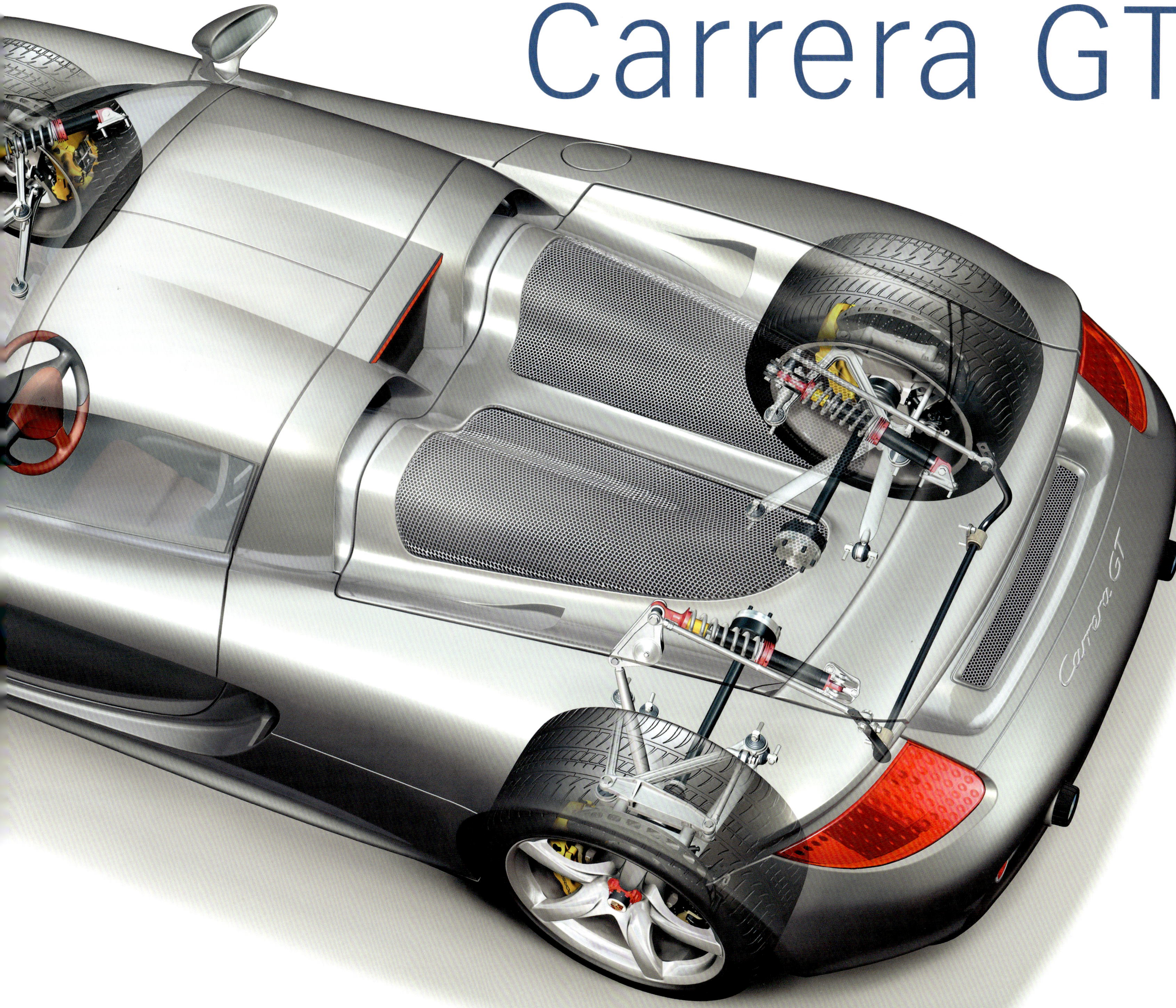

Carrera GT
S GO 612

Carbon fiber chassis and brakes

» Looking at the sleek body of the Carrera GT, it might be thought that it would have a low drag coefficient. But like racing cars, this fast GT did not have a particularly low drag coefficient: it was 0.39. Though only 46 in (1.166 m) high, the Porsche drove on wide tires (at the front 265/35 ZR 19, at the rear 335/30 ZR 20) that undoubtedly contributed some aerodynamic drag. But its aerodynamic features certainly helped the car at high speed. To ensure that the Carrera GT remained stable, without skidding at high speed, it had a complex air deflector underneath that created ground effect.

The spoiler creating downforce on the rear axle was activated automatically at 75 mph (120 km/h) but it could also be released manually. Last but not least, the cooling air vital for the brakes detracted slightly from the aerodynamics.

Considering the price, the interior was rather sparse. At least some of the interior had leather upholstery. The Kevlar-carbon fiber Recaro bucket seats weighed only 22 lb (10 kg) and dominated the interior. Aluminum pedals and a short, handily placed gearshift lever reflected the functional weight and racing-ergonomic concerns of the manufacturer. The two-part roof could be stored in the luggage compartment.

Carrera GT

Porsche 911

A development that would last for decades

» Professor Dr. Ferry Porsche himself wrote that no one would ever have think that the 911 was a development that would last for decades. Today we know better. In the early 1960s Porsche planned the successor for the 356, the development potential of which had been exhausted. The rear air-cooled boxer engine remained unchanged. But notice was taken of the clients' wish for more power and greater smoothness, better roadholding, more space for luggage, and much more. Porsche presented the 901 at the Frankfurt Motor Show in September 1963; it had a body designed by Ferdinand Porsche, the main characteristics of which are still valid today, almost 50 years after it was first presented. The 901 was renamed the 911 after Peugeot claimed that it owned the rights to car names with a zero in the middle. Thanks to Porsche's farsightedness the car, including in particular the two-liter boxer engine with chain-driven overhead camshafts, was conceived with increased performance in mind. Here too the designers laid the foundation for success in racing because the dry sump lubrication and possible high rpm made it easier to construct racing engines. The 911 had long acquired the status of an icon when, about ten years after the start of production, Norbert Schäfer created the first cutaway drawings.

Normal and turbo engines

» In the fall of 1973 Porsche presented a modified 911: the graceful, elegant fenders, smoothly integrated into the body, gave way to massive fenders that were attached seamlessly to the body with rubber bellows. For purists this was the beginning of the end, a derailment of taste, but US legislation required that a car's fenders should remain undamaged when bumped at 3 mph (5 km/h). The length of the car increased from 163.9 in to 168.9 in (4.163 to 4.291 m). The sturdiness of the 911s built from 1976 was also greater. The bodies were made of galvanized steel panels and Porsche was so convinced of the car's corrosion resistance that it gave a 6-year guarantee against rust. The G series of the 911 continued to be developed for many years until 1989. The fifth speed became standard, the engine capacity increased from 165 to 183 cu in (2.7 to 3.0 liters), and later to 195 cu in (3.2 liters). Integrated head supports, automatic belts, and a steering wheel with large hub cushions further increased safety. Exhaust emission controls continued to develop until the three-way catalytic converters became part of the normal specification.

The 911 and 911 S were variants of the first G models with an engine capacity of 164.8 cu in (2.7 liters), for a time the only "everyday" version with a normally-aspirated engine next to the sporty Carrera RS 3.0.

In 1978 the 911 SC 3.0 appeared, with a non-turbo engine. The letters "SC" stood for "Sport Carrera" while the engine had a capacity of 183 cu in (3 liters). Then in the model year 1984 the Carrera designation returned and at the same time the capacity rose to 195.3 cu in (3.2 liters).

911 G

A turbo with 3.0 liters

» After exactly 25 years of activity in Zuffenhausen, Porsche presented the first production turbocharged sports car. In the fall of 1974, a year after the introduction of the bellows fender, Porsche reached new heights with the 911 Turbo. Priced at 65,800 DM, it was very expensive but it also offered the most breathtaking driving performance. The 183-cu in (3.0-liters) engine had remarkable torque – 253 ft lb (343 Nm) at 4,000 rpm – and it was only available with a four-speed gearbox. The 260 bhp gave it a top speed of 149 mph (250 km/h). The large rear wing and the flared rear wheel arches visually distinguished the top model from the 911s with normally-aspirated engines.

With a curb weight of under 2,646 lb (1200 kg) the car had incredible acceleration. The 456-lb (207 kg) engine exerted tremendous tractive force on the rear axle. The fuel consumption was within the wide range of 13 to 17 mpg (14 to 18 liters per 100 km). The first turbos were supplied with 15-inch wheels. In the fall of 1976, the turbo model was given larger 16-inch wheels. The internally ventilated discs with a diameter of 11.1 and 11.4 in (282 and 290 mm) at the front and rear respectively had 2-piston fixed calipers. The brake booster had already been introduced in the previous year. Exactly 2,850 of this first turbo generation were made before the next model change.

911 Turbo 3.0

A turbo with 3.3 liters

›› In 1978 the engine capacity of the Turbo was increased to 201.3 cu in (3,299 cc). In addition, the KKK type 3 turbocharger was now fitted with an intercooler that cooled the compressed air, achieving greater volumetric efficiency. With 300 bhp at 5,500 rpm and torque increased from 250.8 to 303.9 cu in (340 to 412 Nm) the driving performance was even more thrilling – but in the case of the so-called "rest of world" models, clients in Japan, Canada, and the USA had to make do with 265 bhp and 291.3 ft lb (395 Nm). Unlike the non-turbocharged 911s, the Turbo did not have electronic fuel injection (Bosch Motronic) but mechanical manifold injection (Bosch K-Jetronic).

The US/Canada version, and also those intended for Japan and California, differed from the ROW ("Rest Of World") version in their exhaust emission control measures. In addition, this model was fitted with thermal reactors and exhaust gas recirculation to ensure secondary air injection. The Japanese model had grass fire protection while the Californian version was fitted with ignition timing adjustment. Lead-free fuel was essential.

Besides the technical developments prices had also reached new levels. In April 1989 the Turbo coupé cost 138,000 DM while the Targa version cost 146,000 DM, and the Cabriolet an amazing 160,000 DM. In addition to the body illustrated here, Porsche also offered the Turbo with a flat nose. The shape was inspired by that of the successful Group 5 racing-car, the model 935.

911 Turbo 3.3

Fundamental revision

❯❯ After 15 years of model upgrading, the 911 was extensively revised and developed as the 911 type 964.

Externally instantly recognizable by its integrated fenders, the general impression remained unchanged. According to Porsche 85% of the parts were new: the engine, chassis, comfort, and safety features. Initially the new body was only available as the Carrera 4 with all-wheel drive. In 1990 the pure rear-wheel drive made a return.

The last targa in classic form with the removable roof was also derived from the 964 series. Some 93,750 examples of the Carrera 2 and 4, the tuned and weight-optimized Carrera RS, the Turbo, the Turbo S, and the Targa were built up to the fall of 1993 before being replaced by the type 993.

Transition to modernism

» After the long survival of the 911 with its distinctive bellows fenders, 1989 heralded a major technical change in Porsche's approach. In addition to the tried and tested G-models, the Carrera and Turbo, Porsche also developed the Carrera 4 with electronically controlled all-wheel drive. It was an almost completely new car. Instantly recognizable externally with its front and rear skirt, technologically it was also fundamentally different. The engine had a capacity of 219.7 cu in (3,600 cc), the chassis had coil springs all round and a MacPherson front axle. ABS and power steering improved safety and comfort. In 1990 Porsche revised the product range and offered the Carrera 2 as a rear-wheel drive with otherwise identical technology. Compared to the Carrera 3.2, it was about 440 lb (200 kg) heavier, while the all-wheel drive was about 660 lb (300 kg) heavier. The critics were stunned by the road performance of the all-wheel drive. A self-locking differential prevented the rear swerving at the limit of stability.

Carrera 4

Classic rear-wheel drive

» As a result of the new chassis, the rear-wheel drive Carrera 2 was also more easily controllable. Safety was improved thanks to the standard ABS (anti-lock braking system) while the environmental performance was improved by the three-way catalytic converter with Lamda sensor (oxygen probe). The interior was more refined with improved noise suppression. The type 964 therefore offered much greater comfort over long distances. As an alternative to the usual 5-speed gearbox, for the first time a Tiptronic 4-speed automatic transmission was available. As well as the purely automatic mode, the gearchange could also be made by operating the paddles.

Another difference compared to the Carrera 4 was that the Carrera 2 had brake calipers with only two pistons. Also, the type of brake boost was different: in the rear-wheel drive car this was provided by the intake manifold vacuum while the all-wheel drive model had a high-pressure hydraulic system. Other refinements of the year's model: the cabriolet had all-electric soft-top operation, all US vehicles had airbags, the rear seats had three-point seat belts (except in the cabriolet), the headlight range adjustment became standard in production cars, and the speedometer was operated by the wheel impulse of the ABS brakes.

Carrera 2

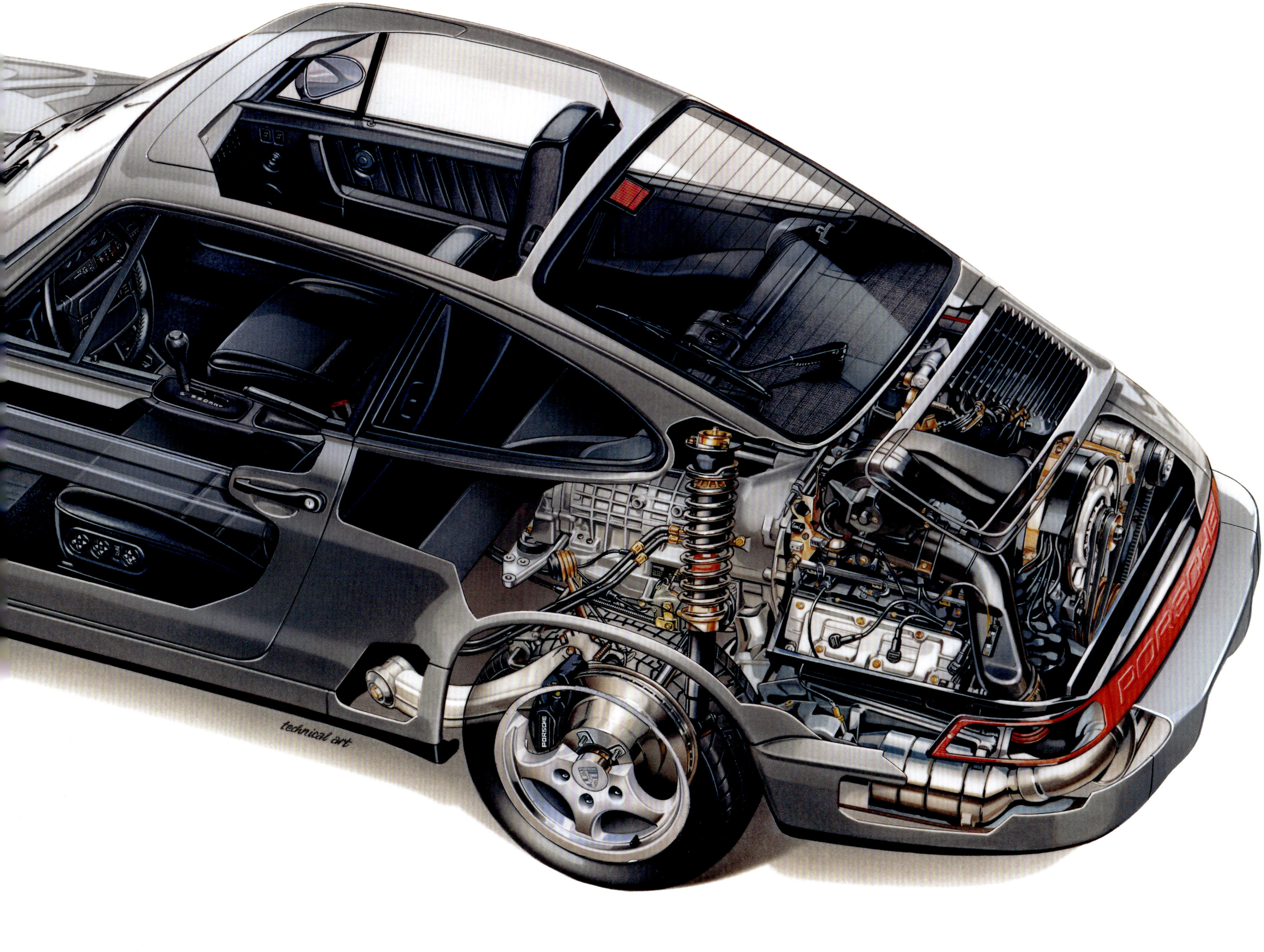

Still 3.3 liters

›› In 1991 Porsche launched another turbo as a top model, again with the famous 201.3-cu in (3.3-liter) engine. The technical engineers had managed to extract a further 20 bhp from the engine, which produced 320 bhp at 5,750 rpm. Also the maximum torque was been increased to 331.9 ft lb (450 Nm) at 4,500 rpm. The body was wider than that of the Carreras. The Turbo was only available as coupé, and only with rear-wheel drive. It had a five-speed gearbox like its predecessor of the previous year. The diameter of the wheels was increased to 17 inches. Reflecting this, the internally ventilated, perforated front brake discs now had a larger friction surface, being 12.7 in (322 mm) in diameter.

In the fall of 1992 the turbo was fitted with the M64 engine with a capacity of 219.7 cu in (3.6 liters). The power output rose further to 360 bhp at 5,500 rpm and the torque was perceptibly higher: 383.5 ft lb (520 Nm) at 4,200 rpm. The wheels increased further in size, now being 19 inches in diameter. The tires were wider: on the front axle they were 225/40 and on the rear axle 265/35. The Turbo 3.6 had an astounding maximum speed of 174 mph (280 km/h). In 1992 the fast turbo look could be specially ordered for the Carrera 2 Cabriolet.

964 Turbo

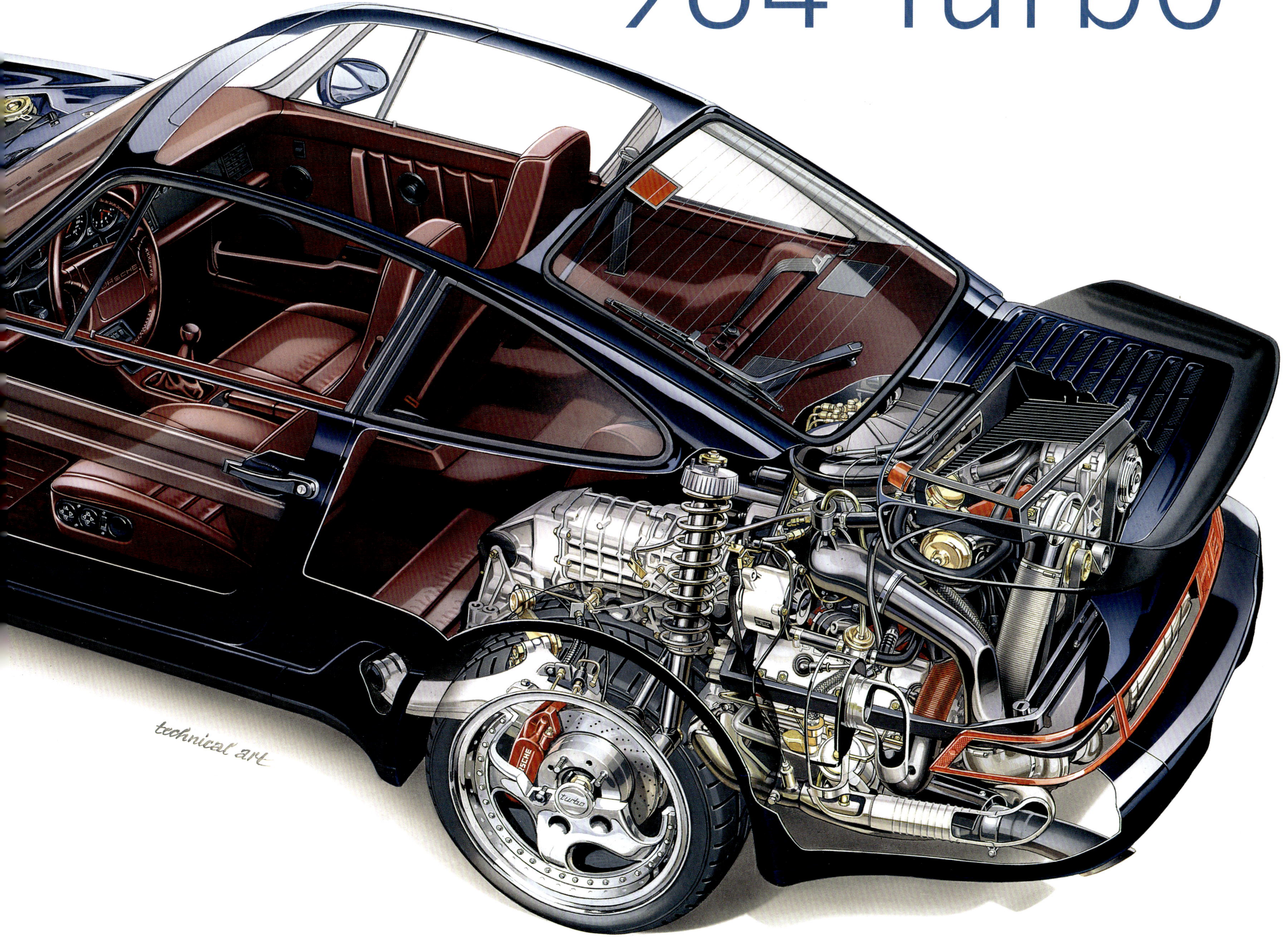

The last air-cooled engine

❯❯ The further development of the 911 type 964 to the type 993 revealed true formal perfection: the fenders were more harmoniously integrated than in its predecessor, and the slightly flatter wings conveyed a particular elegance to the car. In this way the 993 by chief designer Harm Lagaay was reminiscent of the classic 911 built in 1973.

Besides the aesthetic developments, there were also technological modifications. Low-maintenance hydraulic valve tappets for the valve gear reduced running costs. The semi-trailing arm rear axle construction was replaced by a four-arm version known as the LSA, (Light-weight-Stability-Agility) axle. The Carrera 4 launched at the end of 1994 had an all-wheel drive with a viscous coupling unit weighing just 110 lb (50 kg), only half the weight of the earlier system. The Turbo variant was now always fitted with it.

Six-speed gearbox or Tiptronic

» In 1993 Porsche offered the 911 Carrera in coupé, targa, and cabriolet versions. All-wheel drive was available at extra cost. Like its predecessor, the engine had twin-plug ignition, and been given a small increase in power to 272 bhp (in the fall of 1995 it went up to 285 bhp). Because the weight had not increased the driving performance was improved further as a result.

The sporty Carrera RS was more powerful with its 300 bhp, and 220 lb (100 kg) lighter thanks to its spartan fittings; it cost 20,000 DM more. In 1997 it was replaced by the Carrera S with a wider turbo body. But because of its very well-appointed interior, it weighed almost 440 lb (200 kg) more.

The chassis of the 993 type was revised. The front axle with negative steering offset ensured better straight-line stability. The four-arm rear axle made higher acceleration possible, and this was safely transferred to the wider wheels with larger tires. A refined ABS system and perforated brake discs optimized the brakes' performance. The six-speed gearbox with improved synchronization of the two first speeds was also new. The Tiptronic automatic transmission, available as an option, operated more smoothly and was sportier.

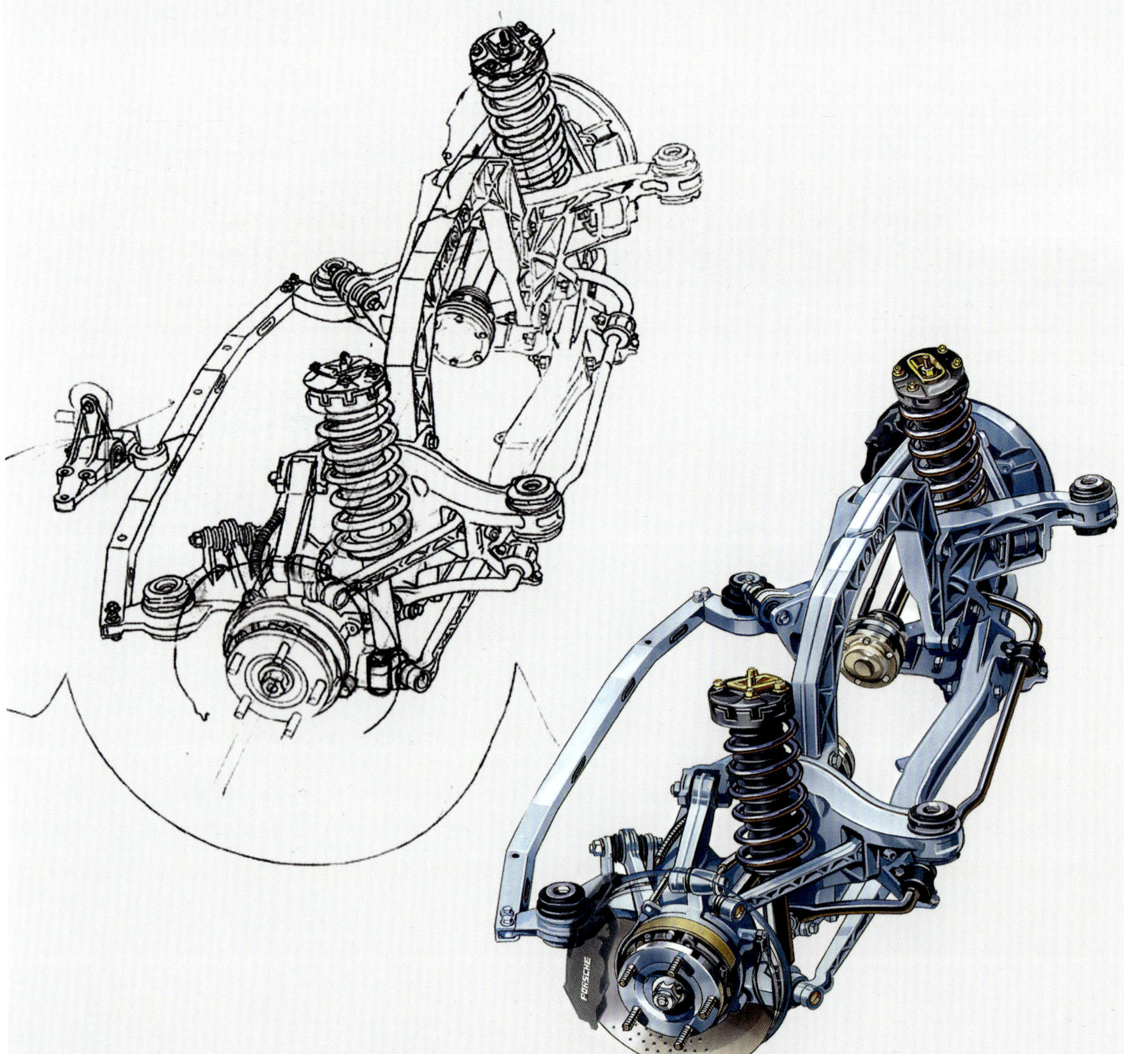

993

With Visco multi-disc clutch

» The 993 Turbo was launched on the market in 1995 as part of the 1996 model year. Its 219.7-cu in (3.6-liter) powerplant was based on the normally aspirated M64/05 engine, and had a forced induction system with two small turbochargers and air-to-air intercoolers. With a compression ratio of 8.0:1, it delivered 408 bhp at 5,750 rpm. The Bosch Motronic M 5.2 system controlled the sequential injection, calculated ignition timing, and determined the boost pressure dependent on air mass. Also new was the OBDII Onboard Diagnostics system, required especially in the USA but used by Porsche in all its turbo vehicles. The transmission of the 993 Turbo was almost the same as that of the Carrera 4, consisting of a six-speed gearbox with modified gear ratios, Visco clutch, transaxle construction, and front differential. The clutch pedal had hydraulic boost diverted from the power steering because the force of 1.5 tons required on the pressure disc for safe power transmission would have made the pedal unreasonably heavy to operate. Besides 6,000 Turbos, about 345 Turbo S cars with 450 bhp were also built between the fall of 1997 and spring 1998.

993 Turbo

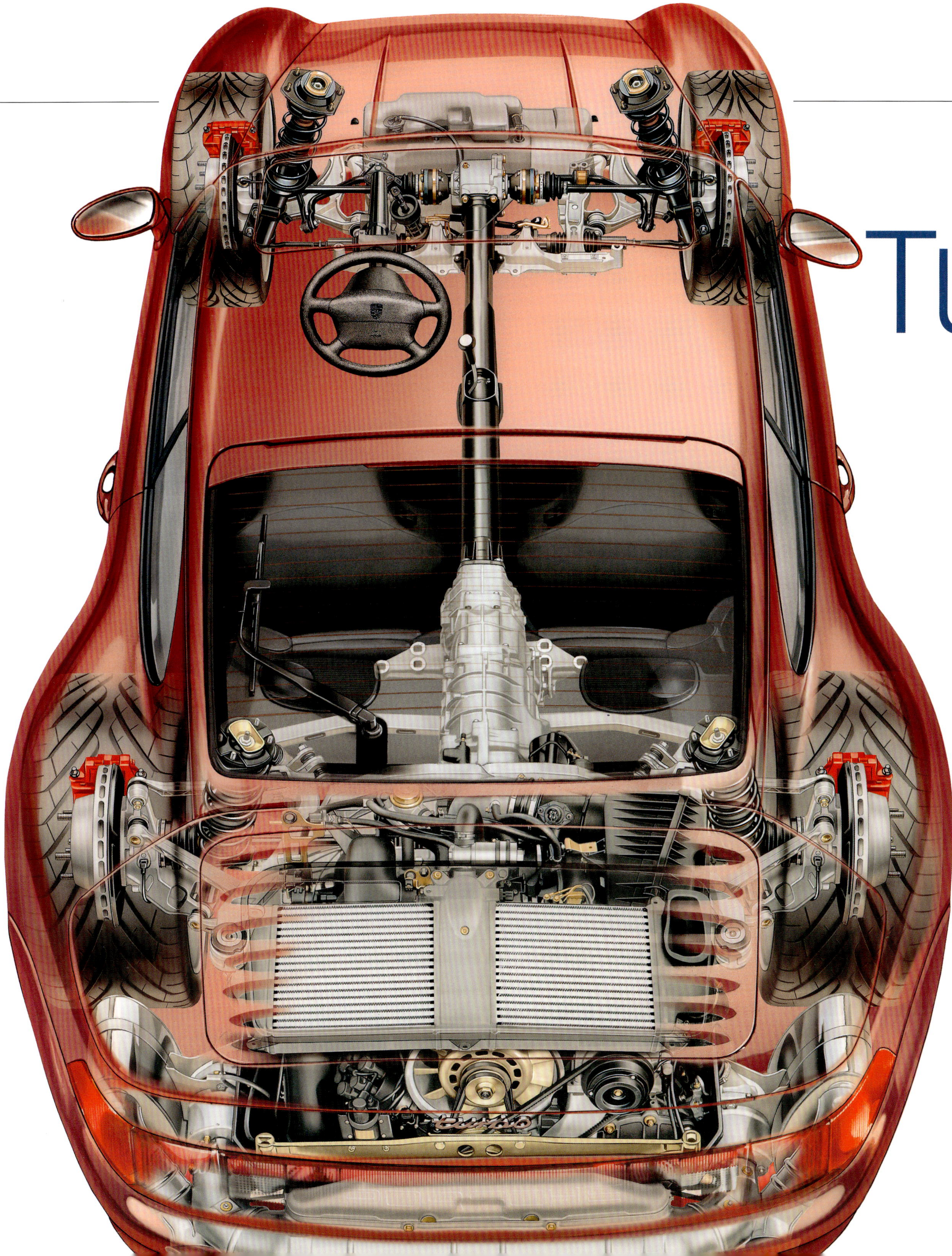

996 – The novelty of water-cooling

›› It could be seen as an upgrade for the mid-engined Boxster, but much more important was the cost-saving "Carry Over Parts" (COP) strategy: visually, the new 911 type 996, presented in the fall of 1997, resembled its little sister. The new 911 looked slightly stretched, and indeed it had a longer wheelbase, by 3.1 in (7.8 cm), it was longer overall, by 7.1 in (18 cm) and slightly wider, by 1.2 in (3 cm). The extra height of 0.2 in (5 mm) was not noticeable. The space in the new car was visibly more generous. Buyers were suspicious that the 996 and the Boxster should be mounted on what was the same body platform from the front to the centre posts. For 911 purists the conversion of the engine to water-cooling was painful. The characteristic sound of the six-cylinder Boxer had changed. Externally, the design of the interconnected cockpit displays was as unpopular as the overall quality of workmanship.

Engine with water-cooling

» The conversion to water-cooled engines was serious. The type M96/01 was 2.76 in (70 mm) shorter and 4.72 in (120 mm) flatter, it had a piston displacement of 206.7 cu in (3,387 cc), and it developed 300 bhp at 6,800 rpm. The ever more stringent noise and emission regulations in particular had made the conversion unavoidable, but also the air-cooled engine could not keep up-to-date with four-valve technology. But for all that, the new engine weighed 110 lb (50 kg) less than that in the type 993.

The seven-bearing crankshaft, unlike the situation with its predecessors, was no longer situated within the engine block but on a separate bearing support. The connecting rods were forged and had a predetermined breaking point; they were assembled separately and adjacent almost without any tolerance. The two overhead camshafts for the exhaust valves were driven by dumb-bell auxiliary shafts that were activated by duplex chains connected to the crankshaft. The VarioCam, known from the Boxster, was used to control the overlapping valve timing for the 996.

996

Axles, wheels, chassis

The further developed 911 type 996 had a body with considerably greater torsional rigidity and flexural resistance. The body was mounted on a completely new chassis. At the front were the MacPherson struts with coil springs and dual gas-pressured shock-absorbers. A longitudinal strut and a crossbar replaced the lower A-frame arm. The turning circle was over 39 in (1 m) smaller compared to its predecessor. Also the rear axle construction was new, designated the LSA-five-arm axle. Four crossbars ensured that the toe-in remained constant, a compensating strut adapting it to all driving conditions. When leaving the factory the 996 had 7J rims at the front, and 9J rims at the rear, the tires being 205/50 ZR 17 and 255/40 ZR 17 respectively. To order, as part of the sports chassis which was lowered 0.4 in (10 mm), or as part of ordinary production from 2001 onward, it was fitted with 225/40 ZR 18, and 265/35 ZR 18 tires on 7.5J rims at the front and 10J at the rear. Four-piston brake calipers built as a monoblock gripped the front internally ventilated perforated brake discs with a diameter of 12.5 in (318 mm) at the front, while the rear ones had a diameter of 11.7 in (298 mm). The ABS was a modern 4-channel system.

In 1998 the Carrera was fitted with a liquid-cooled engine for the first time. The type M96/01 engine, which was 2.76 in (70 mm) shorter and 4.72 in (120 mm) flatter than its immediate predecessor, was fitted to the 993 models with designation M 64/21. Instead of a capacity of 219.7 cu in (3.6 liters), the new engine only had a capacity of 207.5 cu in (3.4 liters). But the power output increased from 285 to 300 bhp, not least as a result of the four-valve technology together with VarioCam. In addition to this high performance technology, the conversion, initially unpopular with purists, ensured that future emission limits could be observed.

The development of the elements in the illustration took place as if in a building set. The images of the individual components were digitally processed and combined.

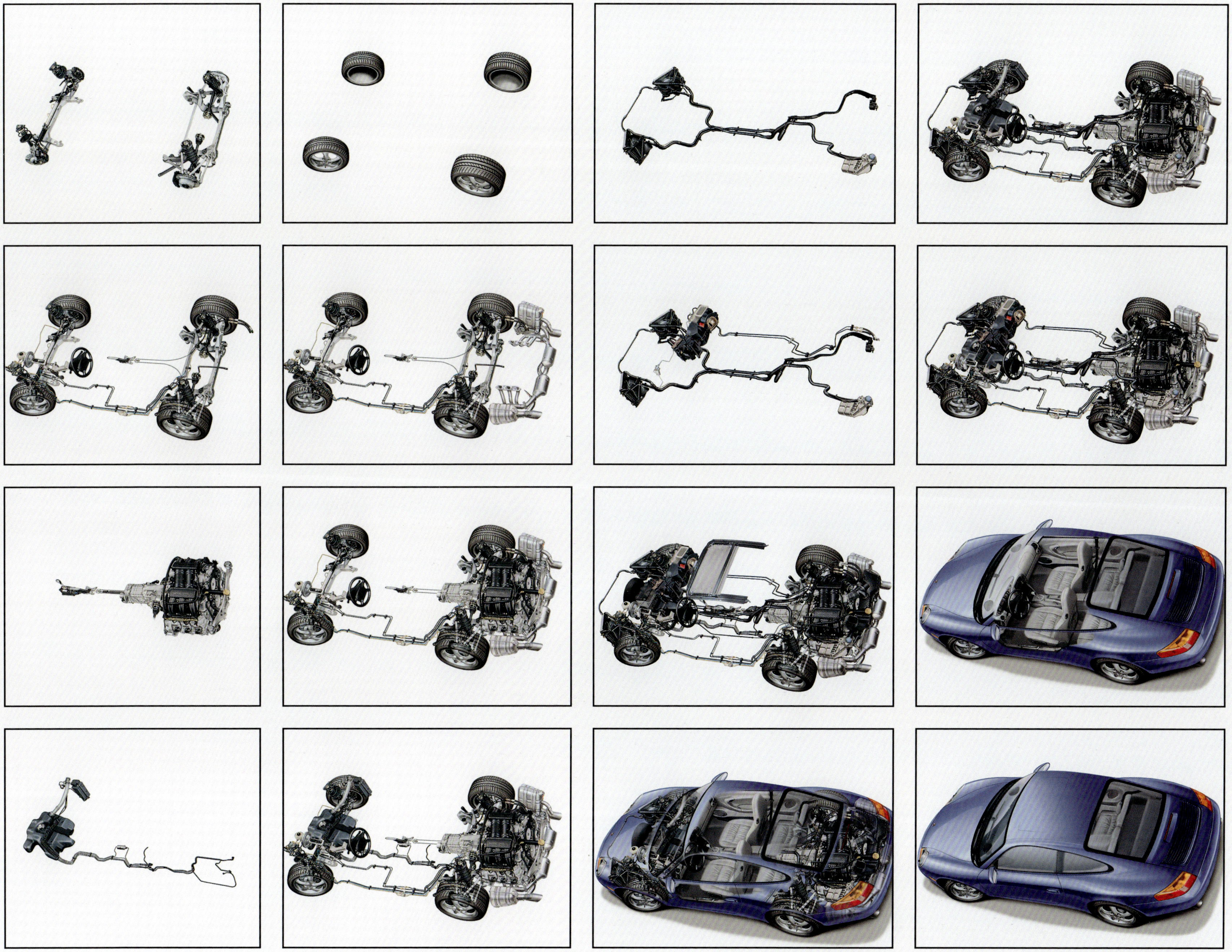

996 C4 – All-wheel drive

›› After the transitional year when the models 993 and 996 were built in parallel, only 911s with water-cooled engines were sold. The new model in the range was the 911 Carrera 4 with permanent all-wheel drive that was available as a cabriolet and as a coupé. The body and engine to a large extent corresponded to the Carrera 2 type 996. A viscous clutch transferred from 5% to 40% of the propulsive power to the front wheels. The Carrera 4 was the first Porsche with an electronic stability program (PSM, Porsche Stability Management).

In dangerous driving conditions the vehicle was stabilized by targeted interventions of the engine management system or the brake system. Because of its electronic stability control, the Carrera 4 was fitted with an electronic accelerator pedal and, unlike the Carrera 2, it no longer had a control cable to operate the throttle.

The information for the throttle actuator was provided by a potentiometer. The all-wheel drive system was also modified compared to its predecessor: the viscous clutch was now contained in the casing of the front differential while the drive shaft between the gearbox at the rear and the front axle differential was uncovered.

Carrera 4

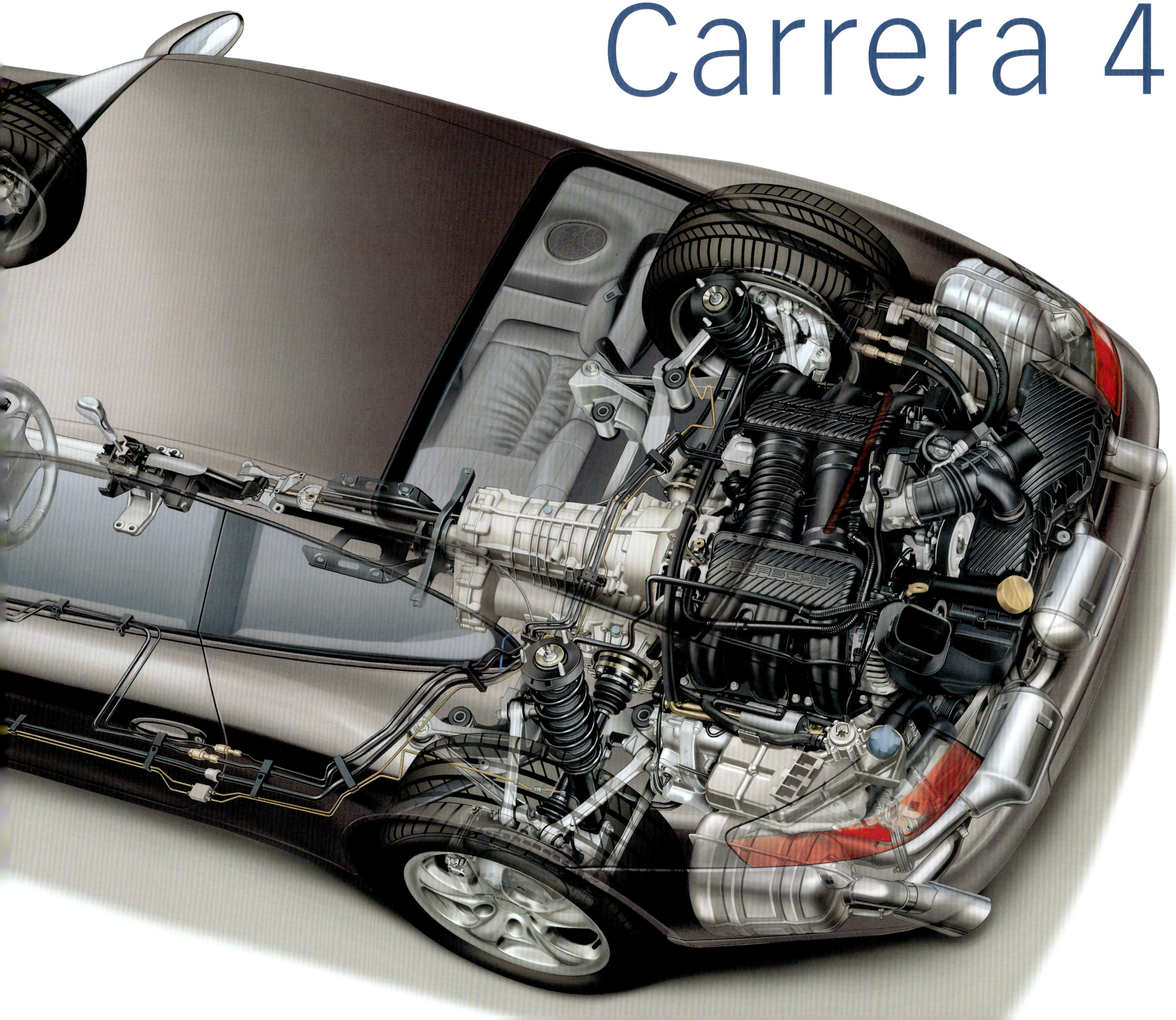

The Turbo's lights and brakes

›› The Carrera 4 S had the visual appearance of the Turbo but the rear of the car housed the normally aspirated engine from the Carrera 2 and 4 combined with the Turbo chassis and brake system. The differences were in the detail: the 4 S had a larger Stabi diameter, a larger master brake cylinder without pressure reducer, and different piston diameters for the brake calipers. The brake discs were larger: at the front 12.99 in (330 mm) instead of 12.52 in (318 mm) in diameter, at the rear 12.99 in (330 mm) instead of 11.7 in (299 mm). They were also thicker, 1.34 in (34 mm) instead of 1.2 in (28 mm) at the front and 1.2 in (28 mm) instead of 0.94 in (24 mm) at the rear. The standard tires at the front were 225/40 ZR 18 on 8J x 18 rims instead of 205/50 ZR 18 on 7J x 17 rims , and at the rear 295/30 ZR 18 on 11J x 18 rims instead of 255/40 ZR 17 on 9J x 17 rims. If punctured, a Tire Mobility System (a repair kit) could be used instead of a spare wheel.
The width of the car was now 82 in (1.830 m) instead of 59.5 in (1.765 m) and the height was 51.0 in (1.295 m) instead 51.4 in (1.305 m). Because of the large tires the track was increased by 0.4 in (10 mm) at the front and by 0.6 in (15 mm) at the rear. Accordingly the curb weight rose by 209 lb (95 kg) and a roof load was not permitted. In spite of the additional weight, the 4 S accelerated one-tenth of second faster to reach 62 mph (100 km/h), and with the Tiptronic transmission it took four-tenths longer (5.6 s). But as far as maximum speed was concerned the Carrera 4 had to admit defeat, and fuel consumption was slightly higher.

Carrera 4 S

Cabriolet concept with safety

» In 1999 the 996 Cabriolet was launched in Carrera 2 and all-wheel drive Carrera 4 versions. The rear-wheel drive car was available by April and the all-wheel version by October. The roof was operated entirely electrically. The thickly lined convertible soft top was completely folded back under a solid cover behind the rear seats by a hydraulic mechanism activated by a switch or the remote control of the central locking mechanism. The process rook about 20 seconds. Reinforcements to the body and the soft top hydraulic mechanism made the Cabriolet about 165 lb (75 kg) heavier than the Coupé.

Also included with the Cabriolet was an aluminum hard top in two parts that, weighing only 70.5 lb (32 kg), was not too difficult to handle and could quickly be installed by two people. This had not been so easy with its predecessor the 993. Safety in the open version has also been improved.

In case the car turned over, two protective roll-over bars immediately rose from behind the backs of the rear seats. They were activated by a tilt sensor and guaranteed head clearance even if the cabriolet ended up with all its wheels in the air.

996 Cabrio

Forever young

» In 2004 the evolution of the Porsche 911 resulted in the type 997 with its more classic appearance. The necessary differentiation with the smaller Boxster ended the uncompromising cost saving COP ("Carry Over Parts") strategy, at least in the case of externally conspicuous features. Initially the new body was fitted to versions of the 911 Carrera and Carrera S coupé as well as the cabriolet.

The model changeover process included numerous changes to the chassis, steering, brakes, wheels, and tires as well as to the upholstery, seats, and interior fittings. Head and side airbags, completely new seats with optimized ergonomics, and much else, improved this slow-burner in the Porsche program. In 2006 the Turbo, GT2, and GT3 were also given the new body.

S GO 134

New generation ...

... with round eyes

» By July 2004 Porsche had already started developing the cars for the 2005 model year. The Carrera, and Carrera S were on the market as the precursors of 997. The headlights of the 996, disrespectfully nicknamed "the fried eggs," returned to their classic round shape, the interconnected instruments were separated again, the classic door handles were used, and many other details reflected the upgrading of the model. The S had a new engine with a capacity increased to 235.4 cu in (3,824 cc), the basic engine having been derived almost completely unchanged from the 219.7-cu in (3.6-liter) one of the 996. In 2006 the Turbo was launched with its new look, and the Targa 4 and 4 S, Carrera and 4 S as well as the GT3 were also given the new look with improved aerodynamics. The G2 had to wait for this until the fall of 2007.

Beautiful, and a 3.8-liter engine

» Besides a price difference of just under 10,000 Euros, the variations between the Carrera 2 and the Carrera 2 S lay in the details. The most significant concerned the engine. In the Carrera 2 the power was provided by the type M96/05 of 219.4 cu in (3,596 cc) and 325 bhp. In the 2 S it was the type M97/01 with 355 bhp. The torque too was different, being 272.9 ft lb (370 Nm) and 295.0 ft lb (400 Nm) respectively. The S-model had marginally more powerful stabilizers and more powerful brakes: the diameter of the brake discs at the front increased from 12.5 to 13.0 in (318 to 330 mm), and at the rear from 11.77 to 13.0 in (299 to 330 mm), and the brake discs in the 2 S were thicker. In addition the 2 S had larger wheels but the 19-inch size was also available on request for the Carrera 2.
Porsche tried hard not to increase the weight of the 997 in relation to its predecessor. In spite of an investment in an aluminum front trunk lid and doors of the same light material, the curb weight increased by 55 lb (25 kg). The maximum speed of 177 mph (285 km/h) remained unchanged but the time from 0 to 62 mph (0 to 100 km/h) was nevertheless two-tenths of second faster.

997 C2

Cabriolet with all-wheel drive

» The Porsche Carrera 4 and 4 S as well as the Cabriolet differed from the Carrera 2 and its variants in that they had all-wheel drive. The connection between the gearbox and the front differential was by a drive shaft. All the wheels are driven by dual-jointed shafts. The additional cost was about 6,000 Euros depending on the model. The front wheels that were also now driven improved driving stability and traction but added an extra 66 lb (30 kg) to the overall weight. This also included the standard oil cooler but the front axle gearbox alone now weighed 51 lb (23 kg). The all-wheel drive model had front brake calipers with six pistons instead of four. The differences in driving performance between the rear-wheel drive car and the all-wheel version were very subtle. Here the all-wheel drive had a slight speed advantage of 3 mph (5 km/h). That was to change when the Carrera 4 S was fitted with a more powerful engine of 381 bhp (instead of 355 bhp), and torque of 306.1 ft lb (415 Nm) instead of 295.0 ft lb (400 Nm).

C4S Cabrio

Coupé and Cabriolet compared

›› The targeted use of body panels of different grades gave the Porsche 911 Coupé great torsional rigidity, and good crash safety. This was also the case with the cabriolet where extra measures had to be taken because of the absence of a roof. In order to maintain the balancing act between adding as little additional weight as possible and achieving the required rigidity, Porsche chose to use high-tensile steel panels and tailored blanks (panels of different thicknesses and various alloys, laser-welded and then compressed together).

With an additional weight of 154 lb (70 kg), the form of the 911 Cabriolet remained the same. The frame of the windshield had been reinforced with high-tensile steel and could serve as a roll-over bar. Additional protection was provided by the roll-over bars that spring out of the centre casing between the back seats and the engine compartment in a fraction of a second if there was a risk of rolling over. Seat belt tensioners and six airbags as well as Porsche Side Impact Protection (POSIP) completed the passive safety arrangements of the car.

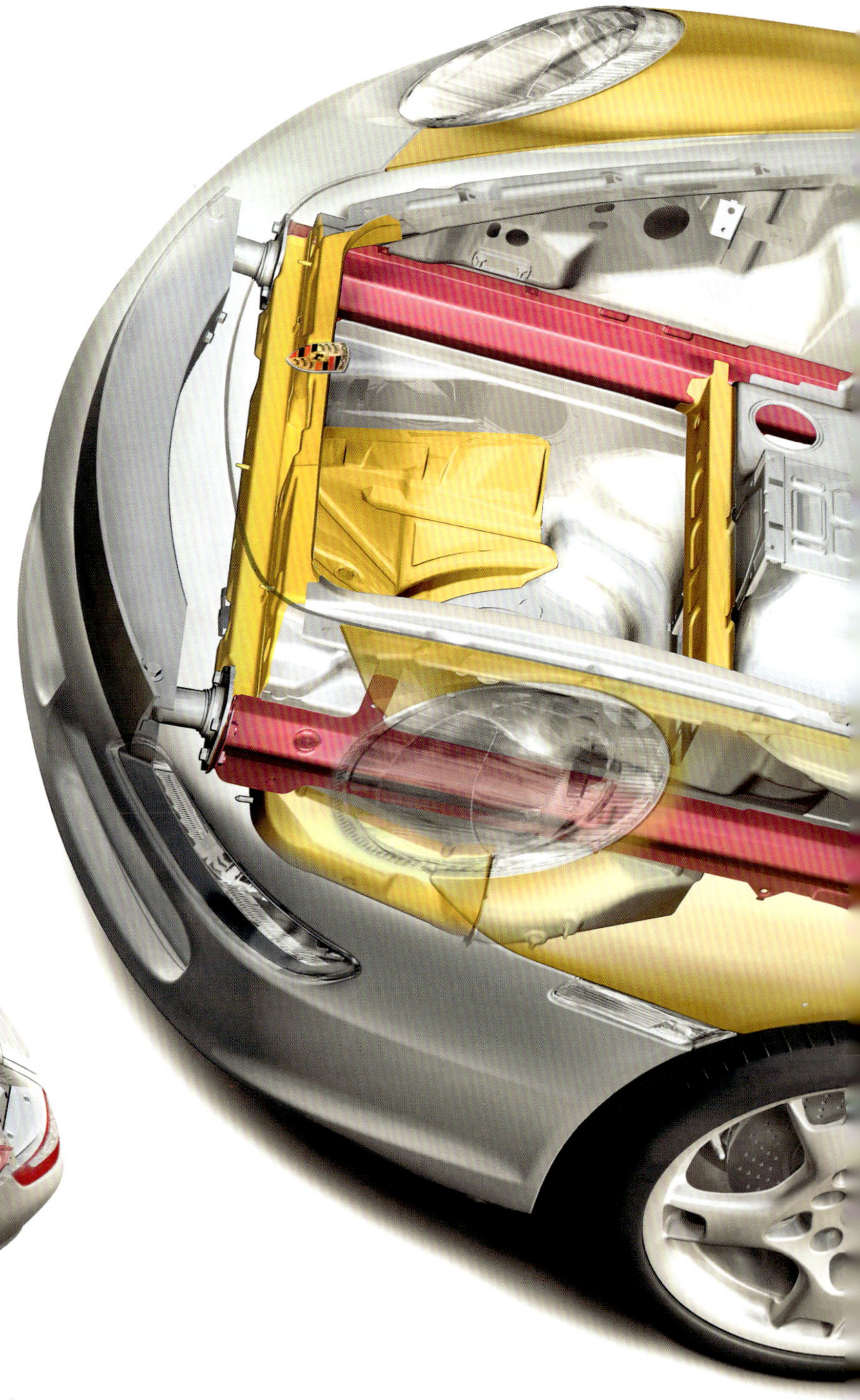

C4S Cabrio

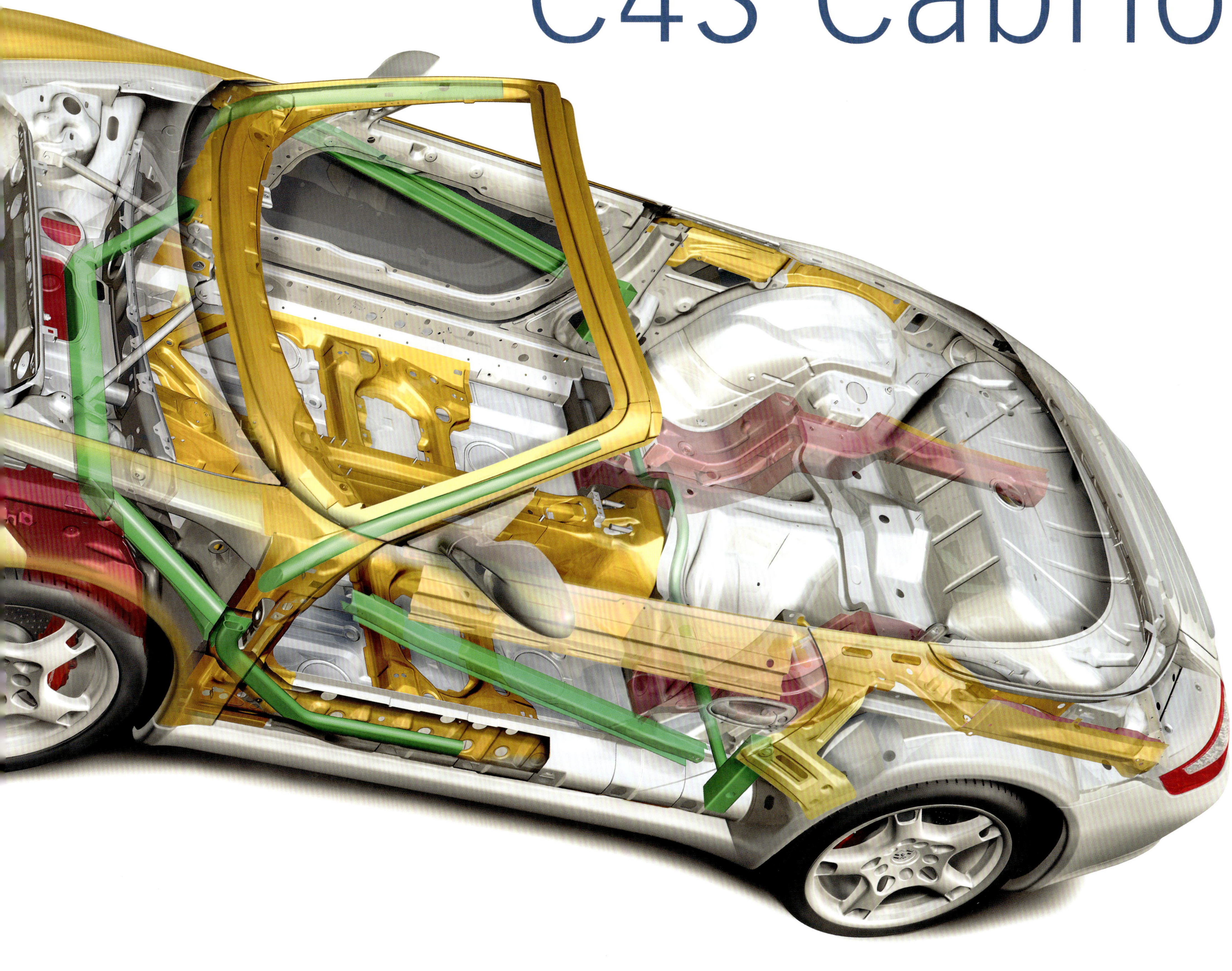

Turbo with multi-disc clutch

» In appearance the Porsche 911 997 Turbo looked quite different from its predecessor in many details. This involved certain features of the body shape introduced in 2004 as well as various changes inside the car. Larger wheels (now 19 inches), modified brakes, a further upgraded stability program (ESP) and the variable steering reduction ratio ensured improved driving safety and comfort.

The all-wheel drive was completely revised. The viscous clutch in the permanent all-wheel drive was replaced by an electronically controlled, electro-magnetically operated multi-disc clutch. This enabled the driver to engage or turn off the front-wheel drive. From 5% to 40% of the propulsive power could be conveyed to the front wheels. Together with the Tiptronic S transmission that shifted without interrupting the tractive power, the Turbo could accelerate to 124 mph (200 km/h) in 12.2 seconds, and it was therefore faster than the manual version (12.8 seconds). In addition, the Tiptronic S had two reverse gears: when starting from cold, a low-ratio reverse gear was engaged, while with a warm engine, the ratio was higher.

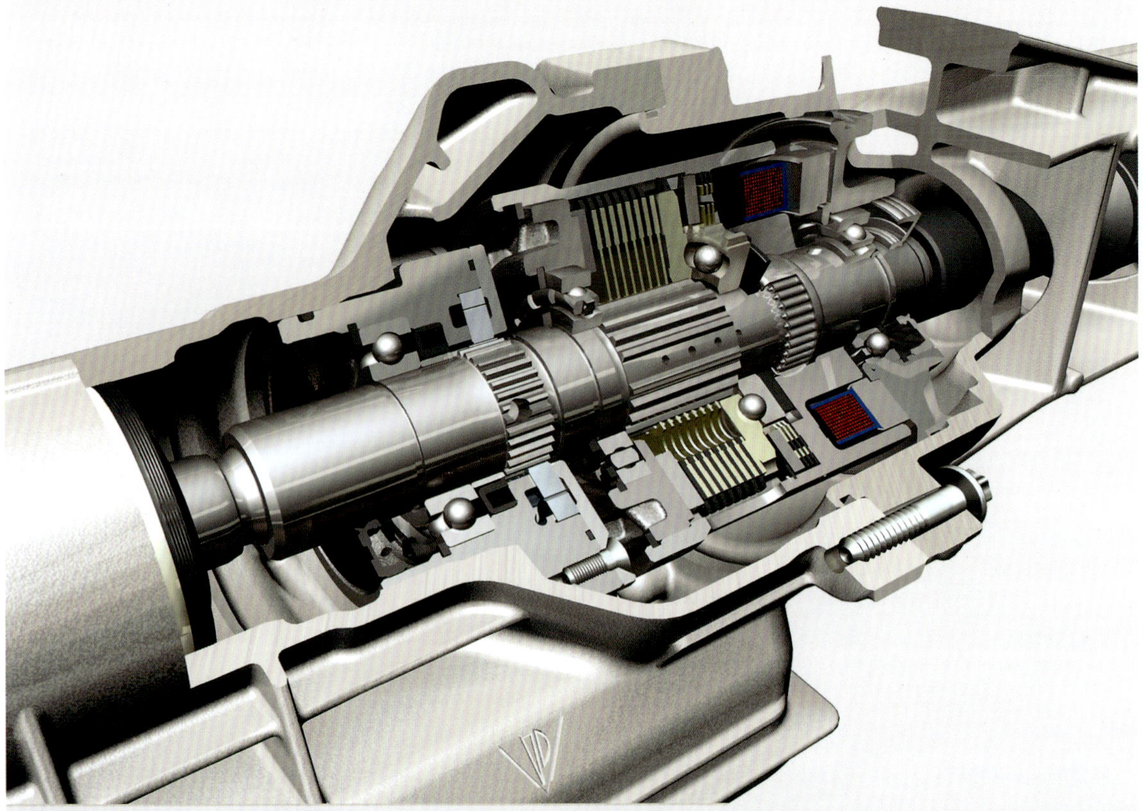

997 Turbo

A complex technology, a compact engine

The 911 Turbo was launched in May 2005 with a list price of 133,603 Euros while the Carrera 2 Coupé cost 76,741 Euros. The enormous difference in price can be explained among other things by the significantly more expensive technology. The bi-turbo engine boasted an extra 155 bhp, thus reaching 480 bhp. There were two Variable Geometry Turbochargers (VGT). Electric control servos changed the angle of attack of the guide vanes according to circumstances.

Thus turbo lag was avoided and the maximum torque of 457.3 ft lb (620 Nm) remained constant between 2,100, and 4,000 rpm. An overboost function was also available at extra cost: the boost pressure could be raised by 0.2 bar for up to ten seconds, producing a torque of 508.9 cu in (680 Nm).

The VarioCam Plus system is also worth mentioning. This enabled the lift of the intake valves to be varied and the timing changed by altering the camshaft position by up to 30° in relation to that of the crankshaft. The significantly reduced fuel consumption values were the result of reduced valve lift at lower rpm, a lift of 0.14 in (3.6 mm) instead of 0.4 in (11 mm), and a second outlet camshaft shape that worked on the exhaust side without overlapping.

Illustrated in the drawing by Technical Art: Air is conveyed to the rear brakes, it cools them, then escapes. As a result the braking action remains fade-free and satisfies the toughest requirements.

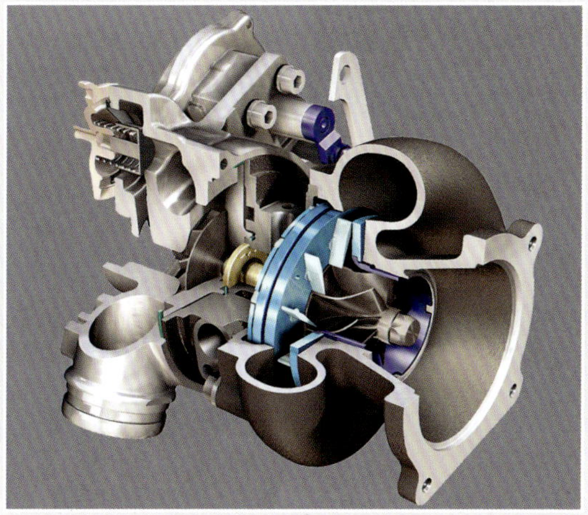

The turbocharger of the exhaust side: these were Variable Geometry Turbochargers (VGT). Electric control servos ensure that the turbochargers adapt to various driving conditions with great precision in spite of an exhaust gas temperature of 1,300 °F (700 °C).

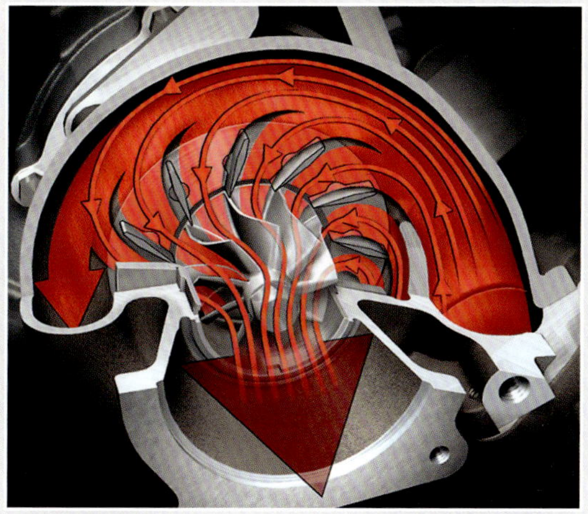

The guide vanes of the turbochargers developed by Porsche in collaboration with Borg-Warner, here seen open, were extremely temperature resistant: in this instance engineers borrowed technology from aerospace engineering. With VGTs turbo lag disappeared and the maximum torque of 457.3 ft lb (620 Nm) remained constant between 2,100 and 4,000 rpm.

A practical roof

» In 2007, besides the 911 Turbo cabriolet, Porsche also introduced a new Targa generation. Since 1967, and up to the type 964, the Targa had always had a removable roof. With the type 993, launched in 2002, it acquired an interesting glass roof construction. As with the cabriolet, additional body reinforcement was necessary and this increased the curb weight by 253 lb (115 kg). The glass part of the roof above the front seats could be electrically retracted.

The safety glass absorbed UV rays. Direct sunlight could be kept out by an electrically controlled blind. The rear part of the roof construction could be opened up like a hatchback, so giving easy access to the storage space behind the front seats. The new Targa was available in Targa 4 and Targa 4S versions, both fitted with the successful all-wheel drive of the 911 Carrera 4. The two versions differed in the price (the 4S was 10,324 Euros more), the tires (18 and 19 inches respectively), and especially in capacity and power output – 219.7 cu in (3.6 liters) and 325 bhp against 231.9 cu in (3.8 liters) and 355 bhp). This explained the slightly better driving performance of the 4S.

C4 Targa

For competition and road use

›› The more sporty version of the Porsche 911 Turbo was the Porsche 911 GT2. Both cars were similar in many respects but the GT2 did not have all-wheel drive and was only available with a six-speed gearbox. The sport version came with an extra charge of 100,000 DM. But in some respects this bought less: no folding seat, no inclinable seat backs, no ESP (Electronic Stability Program), no spare wheel, but most importantly, a curb weight that was 220 lb (100 kg) lower. The long-lasting ceramic brakes were a standard component of the GT2.

S GO 1320

Suitable for competition

When developing the aerodynamics of the GT2 that could be used both on the road and in competition, the Porsche engineers concentrated on avoiding lift force. The redirection of the air flow over the bonnet and the small spoiler lip ensured a good grip of the road. As a result air resistance remained pleasantly low with a drag coefficient of 0.34.

The GT2 version of the 996 pulled ahead with a top speed of 196 mph (315 km/h) instead of 190 mph (305 km/h), and acceleration from 0 to 62 mph (0 to 100 km/h) in 4.1 seconds instead of 4.2, although the Turbo had lower air resistance (0.31). But the GT2 had a smaller frontal area, it was 220 lb (100 kg) lighter, and it was 0.8 in (20 mm) closer to the road.

The aerodynamics of the GT2 were modified as compared to those of the Turbo. On the one hand, the tuned-up model with higher driving performances required more negative lift, while on the other, larger cooling air vents were also needed to maintain the right thermal conditions. Thus the drag coefficient rose from 0.31 to 0.34, although the frontal area was marginally smaller at 21.1 sq ft (1.96 m^2) instead 21.5 sq ft (2.0 m^2).

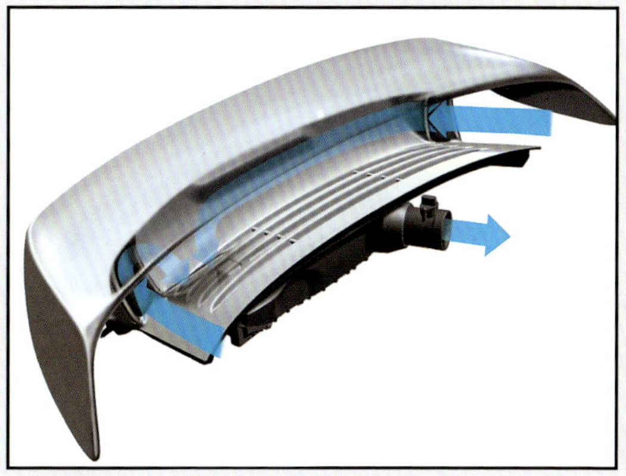

The discreet rear spoiler of the Turbo that would rise up at 75 mph (120 km/h) was replaced in the GT2 by a more prominent, rigid version. But unlike its 993 predecessor, the 996 GT2 was not a homologated model for competition. For sport lovers there was a Club Sport version that was fitted with a bolted-on roll-over cage.

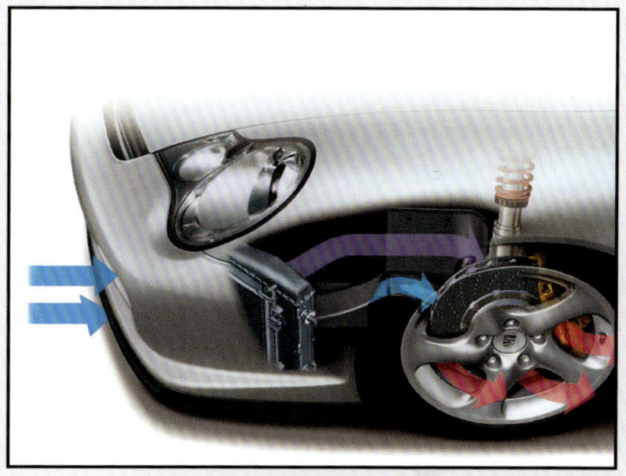

The GT2 was the first production Porsche to be fitted with ceramic brakes (PCCB). The discs at the front and rear had a diameter of 13.78 in (350 mm) and the monoblock brake calipers had six brake pistons. The adjustment of the ABS system was geared to the specific properties of the GT2.

Body structure and powerplant

›› The Porsche GT2 was powered by a tuned-up Turbo engine: two large turbochargers provided air at a pressure of 1.0 bar (Turbo 0.9 bar) to the cylinders, the engine electronics were adjusted and so was the exhaust system. With 462 bhp at 5,700 rpm, it developed 42 bhp more than the Turbo. The maximum torque also increased by 44.3 ft lb (60 Nm) so that in all 457.3 ft lb (620 Nm) was available between 3,500 and 4,500 rpm. In 2003 another big step forward was made: 483 bhp, and 472.0 ft lb (640 Nm). Both the GT2 and Turbo were fitted with the complex valve gear with twin chains, intermediate shafts, and gear wheels as well as the VarioCam Plus. In neutral, the GT2 ticked over at about 740 rpm, slightly slower than the Turbo, but this was certainly not the reason for the slightly lower fuel consumption in city traffic. Both models had the same body structure; in the case of the GT2, part of the front luggage compartment was used to house the larger fuel tank that held 23.5 gallons instead of 17 (89 liters instead of 64). The body was 0.8 in (20 mm) closer to the road so that the GT2 only had a ground clearance of 2.75 in (70 mm).

High-tensile steel
Steel plate
High-strength steel
Boron-alloyed steel
Tailored blanks
Stainless steel

GT2

Porsche 997 as the GT2

» The 911 GT2 built on the basis of the type 997 was presented at Frankfurt in September 2007. With a curb weight of 3,175 lb (1,440 kg), it was 64 lb (29 kg) more than the GT2 built on the basis of the 996. The wheels were now 19 inches in diameter, and the brake discs had also increased in diameter, now being 14.97 in and 13.78 in (380 mm and 350 mm) at front and rear respectively. Air resistance was reduced and the drag coefficient was just 0.32 (previously 0.34).The tire tread width of the new model was also increased, by 0.8 in (20 mm) at the front and by 1.2 in (30 mm) at the rear. In addition, it was even closer to the road surface with a ground clearance of 2.1 in (55 mm) instead of 2.75 in (70 mm), both with the permitted weight. The new engine still had a capacity of 219.7 cu in (3.6 liters) but it was clearly more powerful. The power output had risen from 483 to 530 bhp. The maximum torque had also increased, from 472.0 to 501.5 ft lb (640 to 680 Nm) at 2,200 rpm. A newly developed expansion-aspiration system and variable geometry turbo-chargers contributed to this. For the first time titanium exhaust pipes were fitted to a production car. The GT2 accelerated from a standing start to 62 mph (100 km/h) in 3.7 seconds. The Launch Assistant automatically kept maximum power transferred to the rear wheels at the limit of road grip.

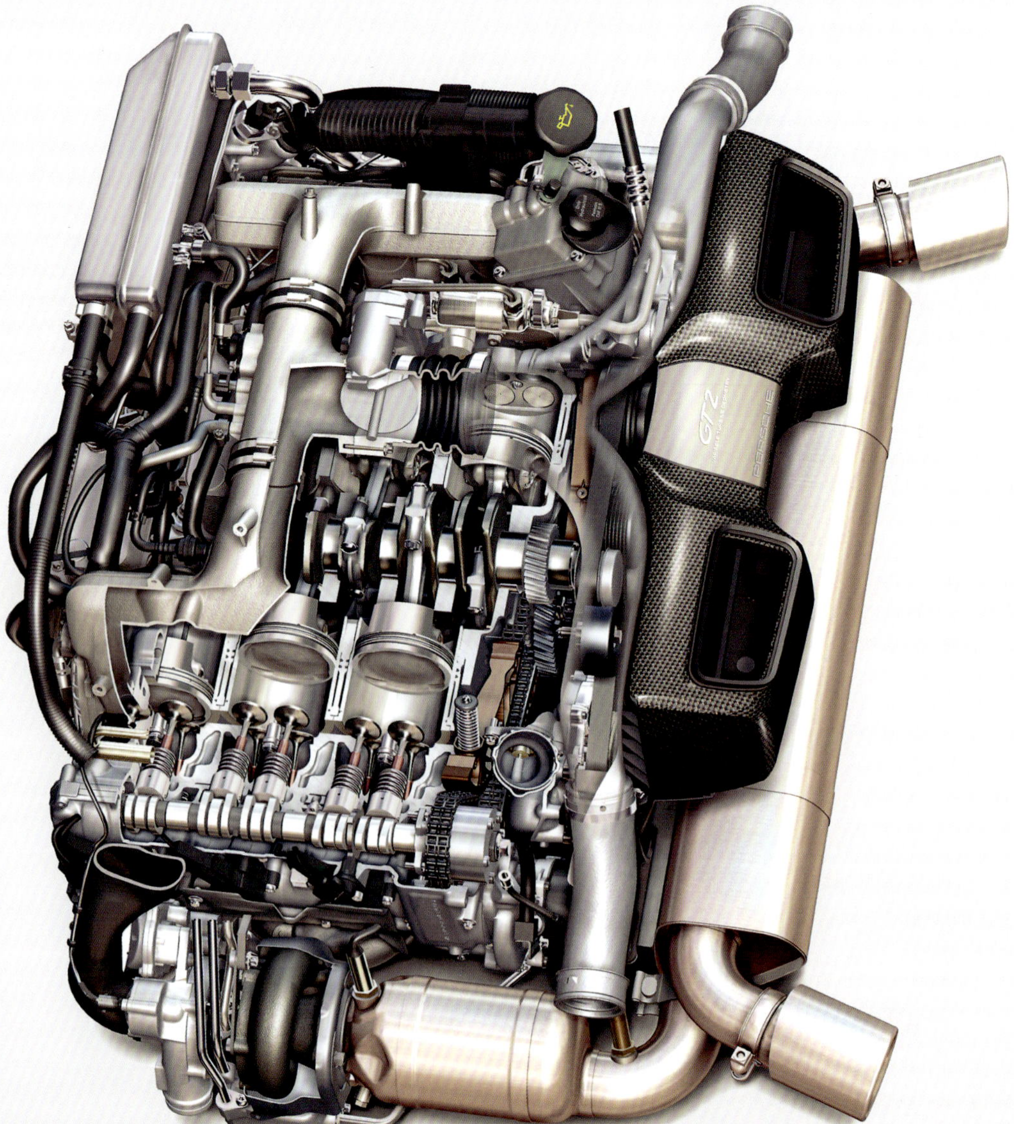

GT2

Inspired by motor sport

›› The letters GT stand for Gran Turismo, but they should not be associated with peaceful touring when they are paired with the figures 911. The model Porsche 911 GT3, based on the type 996, could be used in competition but also on the open road. The added features in the fittings had been reduced to what was indispensable for sports driving. The Club Sport variant also had a bolted-on safety cage. The GT3 was successful both on the race track and in sales, with the second generation already following in 2003, and a lighter RS-version was also built. In 2006 Porsche launched the GT3, built on the basis of the type 997, an RS version of which was also available that already included the Club Sport package.

The Cayenne paved the way, but the GT3 also came into being at Technical Art in 2002 through a new technique: for the first time 3D data were used for the illustrations.

GO 2156

The fastest non-turbo engine

›› The chassis of the GT3, basically derived from the type 996, was partly modified. It was set 1.2 in (30 mm) lower, had adjustable stabilizers, and height-adjustable spring plates on both axles.
The pressurized shock-absorbers were set for sports use while larger, thicker brake discs took into account the increased power output. The six-speed gearbox type G96/90 was just over 17.6 lb (8 kg) heavier than the Carrera 4 (G96/30). The separate oil tank was in keeping with its use in competition. Visually, the model had conspicuous front and rear sections as well as an eye-catching spoiler, but these features could also be seen on other 911s because they could be ordered as extras.

The engine of the GT3 was very different from that of the Carrera engine. With a capacity of 219.7 cu in (3.6 liters), it developed 360 bhp. With its larger bore and shorter stroke, eight-bearing optimized crankshaft, titanium piston rods, and wet cylinder liners, it could reach up to 7,800 rpm. The VarioCam camshaft with 25° adjustment of the induction camshaft operated the valves with twin valve springs and hydraulic valve lift adjustment in the six individual cylinder heads. The increased performance was also the result of the compression ratio being increased to 11.7:1, and the two-stage resonance-intake was made of aluminum.
The drawing was made conventionally with digital image processing.

GT3

The GT3 evolution of the 997

›› The first series of the GT3 had already been very successful and Porsche had built more cars than originally planned. As a result it launched a revised version in April 2003. The new front with headlights borrowed from the Turbo and a completely new rear wing distinguished it visually from its predecessor. With a drag coefficient of 0.30, a maximum speed of 190 mph (306 km/h) was now possible. Porsche had further improved the brake system with 13.8 in (350 mm) brake discs with six-piston fixed calipers at the front, and 13.0 in (330 mm) brake discs with four-piston brake calipers at the rear. Ceramic brakes were available at extra charge.

The engine was given type number M96/79. With further fine-tuning, the power output rose by 21 bhp to 381 bhp at 7,400 rpm. The maximum rpm was 8,200 rpm, 400 rpm higher. The VarioCam Plus enabled the intake camshaft to be adjusted by 45°. The maximum torque rose from 250.8 to 284.0 ft lb (370 to 385 Nm) at 5,000 rpm. The tires were slightly larger too. At the front they measured 235/40 ZR 18 (instead of 225), at the rear 295/30 ZR 18 (instead of 285). The car only weighed 22 lb (10 kg) more than its predecessor. The air-conditioning system was available at no extra charge.
The illustration below was produced from 3D design data.

GT3

911 as the 997 GT3 RS

» In 2004 Porsche launched the GT3 RS in a limited edition. This competition car had to conform with the legal regulations allowing it to drive on public roads, and at the same time meet the requirements of the FIA N-GT class regulations for Europe, and ACO for North America. Externally the RS was identifiable by the three ventilation slits above the front skirt, the rear view mirror bodies in visible carbon fiber, and the projecting rear spoiler. It was 44 lb (20 kg) lighter than the GT3.

In May 2004 the GT3 RS made its debut with a list price of 120,788 Euros, and just 678 were built. By 2005 Porsche had already launched the new GT3 RS on the basis of type 997, and two pilot-production cars were built. In May 2005 the list price for the car was 108,083 Euros; only nine were built. In August 2006 the list price of the car rose to 129,659 Euros, and Porsche was able to build 1,095 of this model. Reinterpreted as GT RSR with a modified kinetic chassis and sequential to manual transmission, it would win in its class in 24-hour races, and as a result was supplied to racing teams throughout the world.

GT3 RS

S GO 1004
S GO 1053
S GO 171
S GO 118

Modern models

›› The large V8-GT with the designation 928 did not replace the 911, nor was the four-cylinder 968 particularly successful either. In the mid-1990s Porsche felt the need to initiate a new project and was on course for success with the Boxster mounted on the 911 body. Part of the production was moved from Zuffenhausen to Finland. Typically Porsche has continuously improved and expanded the range of models. In 1999 the Boxster S was launched and in 2005 a coupé based on the Boxster came out with the snappy name of Cayman S. The following year the range was extended further with an economy model with a smaller engine.

Even before the introduction of the Cayman, Porsche had ventured into sector of Sports Utility Vehicles (SUV). The Porsche Cayenne in the S/GTS and Turbo variants, derived from the Volkswagen Touareg, were already very successful on the market so that malicious tongues began to speak of Porsche as an SUV manufacturer that also made sports cars. Porsche has developed several ver-sions of this model as it has always done with others, so there is now also a Porsche Cayenne with a diesel engine.

The current Porsche 911 models are included among the classics in spite of the title of this chapter.

Porsche Boxster

A successful mid-engine roadster

» **In January 1993**, at the North American International Motorshow in Detroit, Porsche presented the prototype of a new mid-engine roadster. The American market is very important for the Zuffenhausem manufacturer and the response was positive. It was not until August 1996 that the Boxster, the name of the youngest Porsche scion, was launched on the Ger-man market.

The model owed its name to the contraction of the name of its engine, "boxer," and the word "roadster." Compared to the prototype, only small details such as changes to the air inlets at the front and on the sides were apparent.

The Boxster was convincing both visually and technically. With a few borrowings from its big brother 911 that at the same time led to synergies as a result of the "carry over parts" concept, the Boxster was immediately well received. In fact, it had a considerably better start than the 914 and 924 models had had. The relatively modest starting price of just over 76,000 DM – just over half the price of a 911 cabriolet – also contributed to this success. The Boxster achieved record sales for Porsche. Because the high demand led to a bottleneck in its production capacity in Zuffenhausen, part of the production was moved to Finland.

Since 1997 the Boxster has continued to be built in parallel there by Valmet Automotive in Uusikaupunki.

The Porsche 986 named Boxster

›› The in-house code for the Boxster, built since 1996, was 986. It was 170.4 in (4.329 m) long and had a wheelbase of 95.1 in (2.415 m). At the push of a button, the electrically-operated roof folded back directly behind the seats into a container mounted above the mid-engine. There was sufficient storage apace under both the hood at the front and the trunk at the rear. The self-supporting steel body had a retractable spoiler at the rear. MacPherson strut suspension at front and rear with stabilizers were in keeping with the latest technology, as was the brake system with internally ventilated steel discs and ABS. The 2,756-lb (1,250-kg) Boxster had 16-inch tires, 205/55 at the front and 225/50 at the rear. The liquid-cooled six-cylinder Boxer engine had a capacity of 151.3 cu in (2,480 cc) and developed 204 bhp at 6,000 rpm. The maximum torque of 180.7 ft lb (245 Nm) was reached at 4,500 rpm. Two overhead camshafts per cylinder bank operated four valves per combustion chamber. The power was transmitted to the rear wheels by a manual five-speed gearbox or by Tiptronic automatic transmission. With a maximum speed of 149 mph (240 km/h), and an acceleration of 0-100 km/h in 6.9 seconds (7.6 seconds with Tiptronic), excellent performance was achieved.

The illustrations were produced conventionally way with subsequent digital image processing.

Boxster

The "S" version with 3.2 liters

» Continuous model upgrading and a broadening of the range offered kept the demand for the Boxster high. For the 2000 model year, Porsche increased the engine capacity of the economy model. An extra 0.24 in (6 mm) stroke increased the capacity to 164.0 cu in (2,687 cc). The power output rose to 220 bhp at 6,400 rpm and the maximum torque was now 191.8 ft lb (260 Nm) at 4,750 rpm. In the model year 2003 Porsche introduced the adjustable "VarioCam" camshaft, and the power output increased to 228 bhp at 6,300 rpm. Fortunately the Boxster did not put on any weight worth mentioning – just 22 lb (10 kg). The maximum speed was 157 mph (253 km/h).

The Boxster S that also came out in the model year 2000 was even more rakish. The rebored engine with a capacity of 194.0 cu in (3,179 cc) developed 252 bhp at 6,250 rpm and also had a higher torque of 225.0 ft lb (305 Nm) at 4,500 rpm; with VarioCam it was 260 bhp at 6,200 rpm, and 228.6 ft lb (310 Nm) at 4,600 rpm. Besides the five-speed Tiptronic automatic transmission, a manual six-speed transmission was also available. In spite of an curb weight of just under 1,2,866 lb (300 kg) the Boxster S still had a maximum speed of 162 mph (260 km/h) and accelerated from 0 to 62 mph (0 to 100 km/h) in a little under six seconds. Connoisseurs would immediately recognize the Boxster S by its large tires. On the rear axles, massive 17-inch 255/40 tires clawed the tarmac, while 17-inch 205/50 ZR tires were fitted at the front.

154-155 A successful combination:
the little Porsche sold very well.
The cutaway illustration was made from
3D data combined with photographic elements.

Boxster S

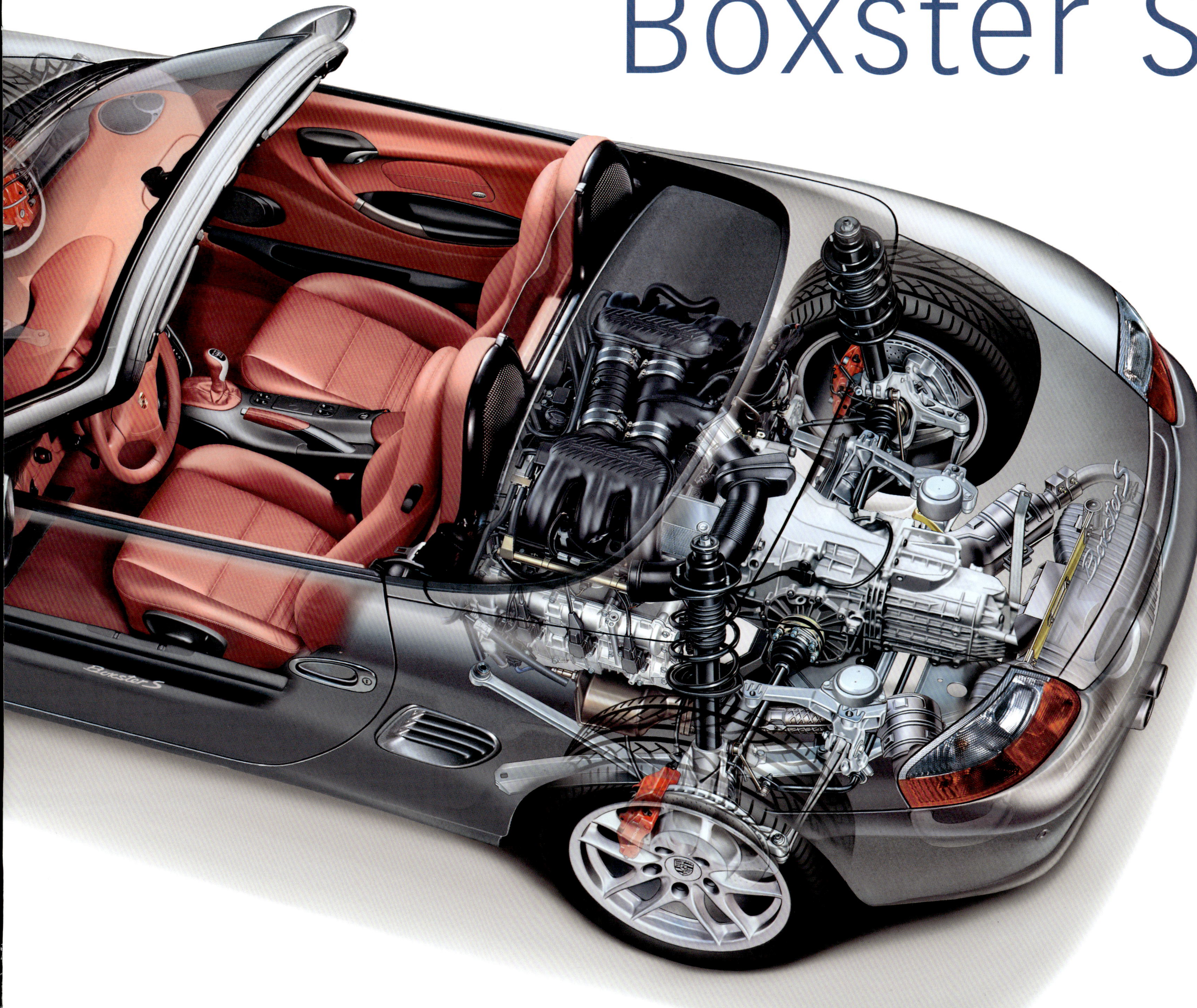

S AZ 356

Versions 2.0 and 2.1

›› After almost 110,000 Boxster 986s had been built, Porsche decided to switch production to the type 987. This arrived on the market at the end of November 2004 and its new headlights clearly distinguished it from its predecessor. The interior has also undergone a facelift with a modified central console, new door upholstery, and a different steering-wheel. The folding top control that made it possible to open and close the roof up to 50 km/h is also new. The basic power unit was still the 164.8-cu in (2.7-liter) engine with a power output of 240 bhp, rising to 245 bhp in model year 2007.
Besides the standard five-speed transmission and Tiptronic, the entry model was also available with a sport package to order. This included a six-speed transmission and the Porsche Active Suspension Management system (PASM) that adjusted the spring characteristics of each wheel. The S-variant developed 280 bhp and in the model year 2007 this evolved into the 207.5-cu in (3.4-liter) model, based on the Cayman, with power output been increased to 295 bhp. From February 2009 onward the Boxster and Boxster S had a new engines with capacities of 177.0 and 207.5 cu in (2.9 and 3.4 liters) developing 225 and 310 bhp, with direct fuel injection and a modified oil supply.

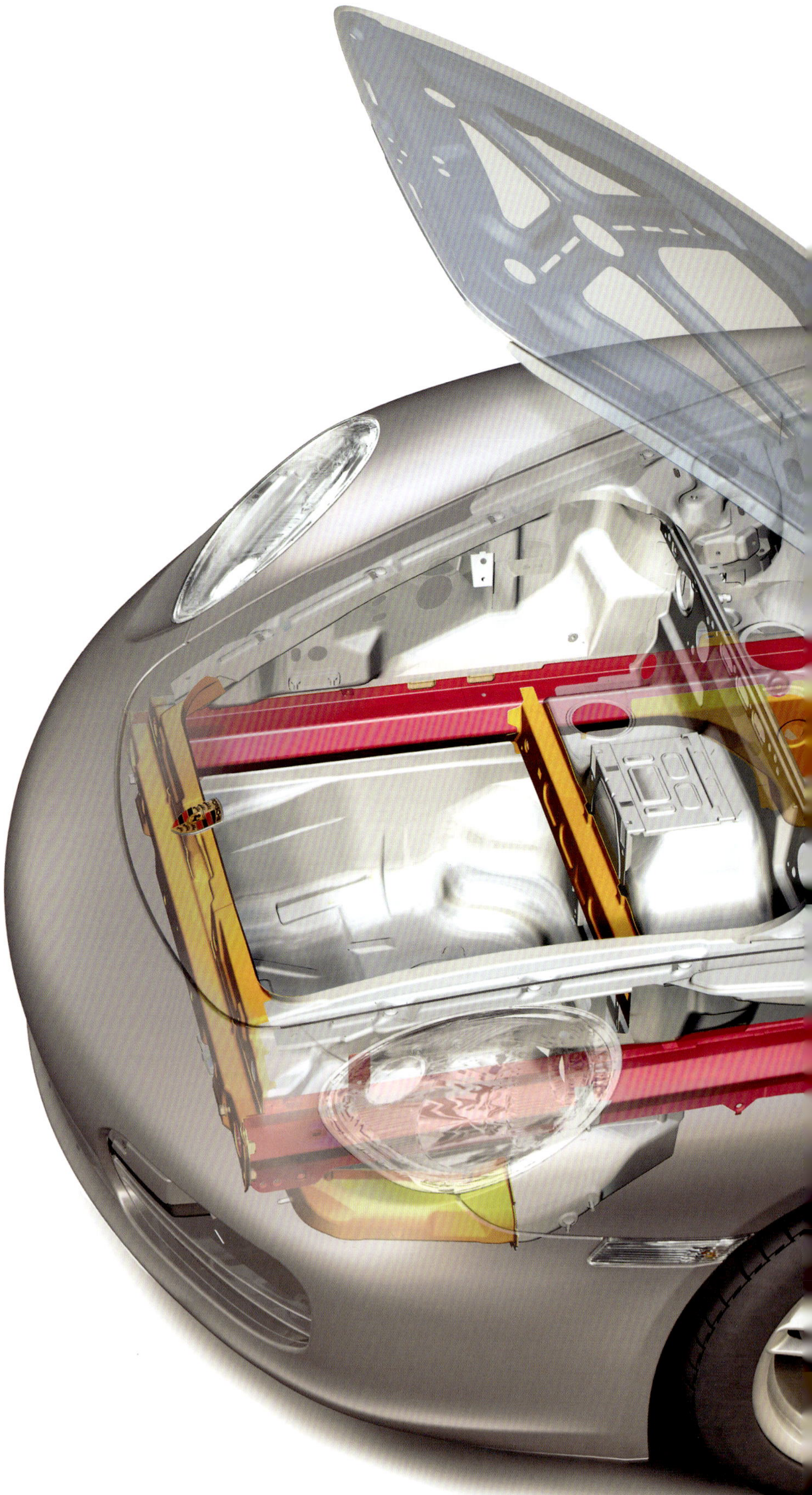

Boxster 987

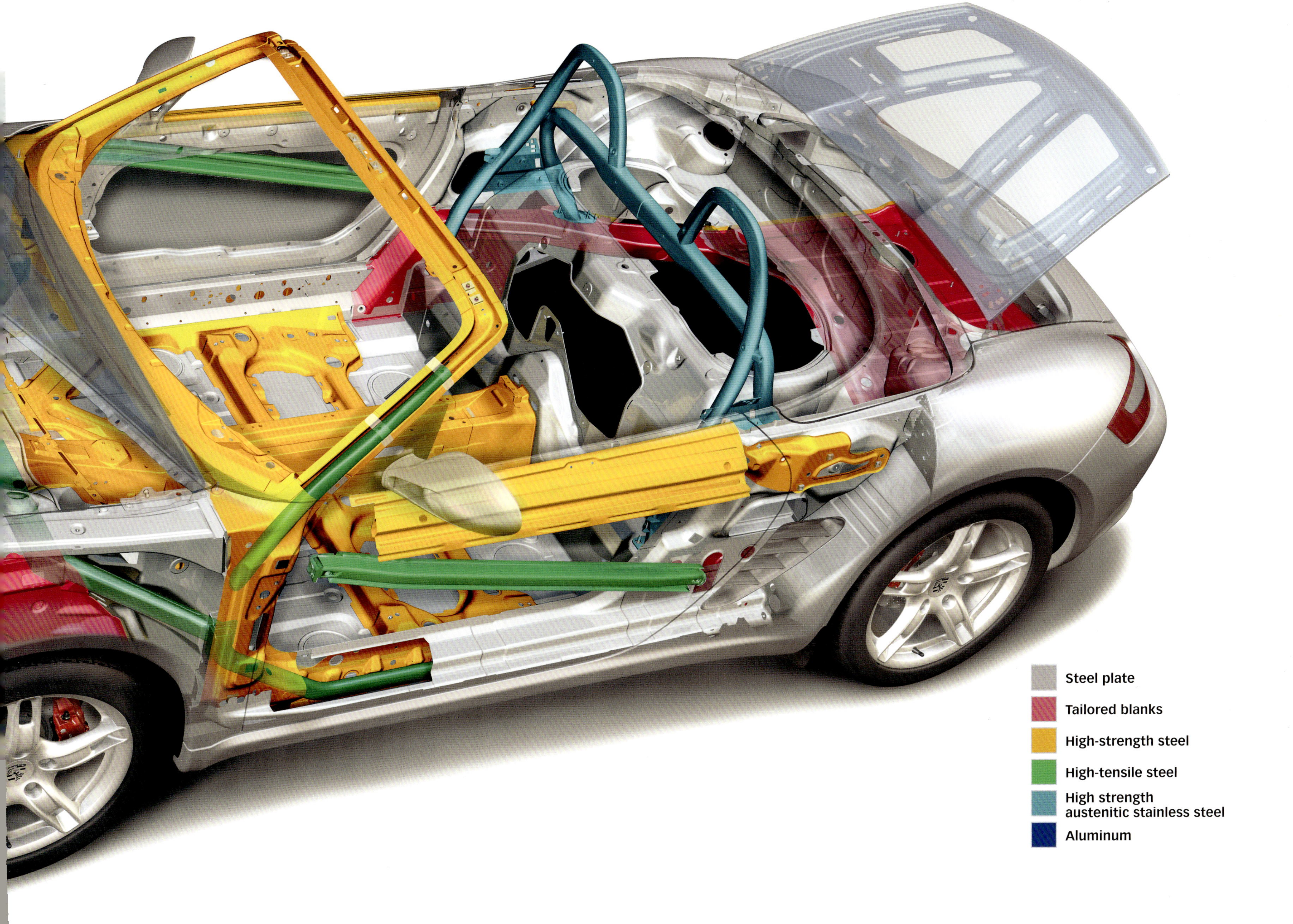
Steel plate
Tailored blanks
High-strength steel
High-tensile steel
High strength austenitic stainless steel
Aluminum

The new Boxster

Mid-engine roadster by Porsche

» **In mid-April 2012** Porsche launched a completely new model of the Boxster. The two-seater convertible has an entirely new lightweight body and a completely revised chassis. A much lower weight, longer wheelbase, wider track, and larger wheels significantly improves the driving dynamics of the mid-engine sports car, these being further enhanced by the electro-mechanical power-steering. In spite of its improved performance, the new Boxster is 15% more economical. Depending on the model, the fuel consumption is considerably less than 29 mpg (8 liters per 100 km).

The modified design with shorter overhangs, steeply slanting windshield, flatter silhouette, and crisp edges confirms the increased performance of the new Boxster. The occupants are protected by the fully electrically-controlled roof. The car's interior provides more space for the occupants and the raised center console is more ergonomic.

Optimum acceleration

›› The new sports car generation made its debut as the Boxster and Boxster S. Both models are powered 6-cylinder engines with direct fuel injection, the efficiency of which has been increased by the kinetic energy recovery system, thermo-management and Start-Stop function. With a capacity of 164.8 cu in (2.7 liters), the new engine of the basic model develops 265 bhp, 10 bhp more than its predecessor that had a larger capacity. Technically it is based on the 207.5-cu in (3.4 liter) engine of the Boxster S.

The S model now develops 315 bhp, that is, 5 bhp more than before. Both models come with a standard manual 6-speed transmission and the Porsche 7-speed twin-clutch transmission (TCT) is available as an option. Both sports cars achieve their optimum fuel con-sumption and acceleration when the TCT system is fitted. The Boxster's fuel consumption is 30.7 mpg (7.7 liters per 100 km), and that of the Boxster S 29.4 mpg (8.0 liters per 100 km). The Boxster accelerates from 0 to 62 mph (0 to 100 km/h) in 5.7 seconds, and the Boxster S in 5.0 seconds. In order to increase the performance further, Porsche offers the Sport Chrono Package as an option; for the first time this includes dynamic transmission range. Also new for the Boxster is the optional Porsche Torque Vectoring (PTV) with mechanical rear-axle-differential lock. The base prices in Germany are 48,291 Euros for the Boxster and 59,120 Euros for the Boxster S.

The new Boxster

Porsche Cayman

The racy coupé

» **At the Frankfurt Motor Show** in 2005, Porsche unveiled a new model. Based on the Boxster, the Cayman S coupé variant was designated 987c. Because the container for a convertible top was no longer needed, there was an additional 9 cu ft (260 liters) of luggage space under the front hood that could be used as a trunk. With a starting price of more than 58,000 Euros it was several thousand Euros more expensive than the Boxster and, like the open model, it was built by Valmet Automotive in Finland. The car came on the market in Germany in November 2005.

Contrary to what is usually the case, this closed version cost more than the cabriolet, the Boxster that had been introduced in 1996. In 2006, after the Cayman S, the slightly tamer version without an "S" was launched. It was 3,000 Euros more expensive than the open Boxster but this did not appear to affect its popularity on the market.

Cayman S

A successful extension program

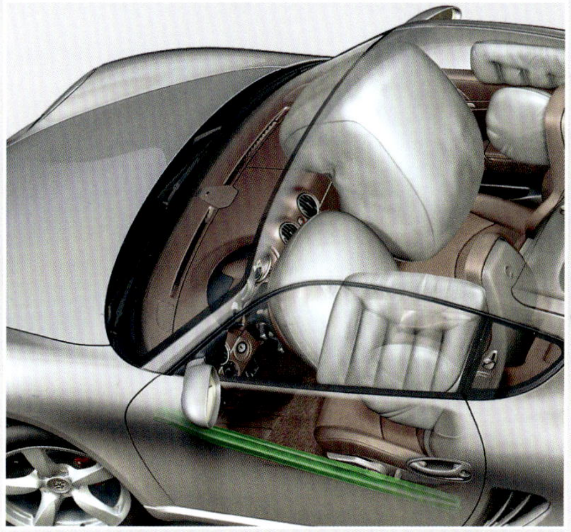

» The sleek Cayman S anticipated the 207.5-cu in (3.4-liter) 6-cylinder boxer engine that only came out in the Boxster version in 2006. In July Porsche launched the Cayman – with no "S" – with a 174.76-cu in (2.7-liter) power unit but with power reduced by 50 bhp. The car's handling was sportier than that of its open sister-model because of its rigid body. Externally the Cayman and the Cayman S differed in that one had a single exhaust pipe and the other a twin exhaust, one had black anodized brake calipers, the other red, one had 17-inch and the other 18-inch tires, and one had a separating edge at the rear and the other a retractable wing. In 2007 Porsche launched a special series of 777 cars of the Cayman S Porsche Design Edition and in 2008 the Cayman S Sport came out with 8 more bhp in a limited edition of 700 cars. After a facelift in spring 2009, the capacities were 177.0 cu in (2.9 liters) and 207.5 cu in (3.4 liters), with greater power and lower fuel consumption.

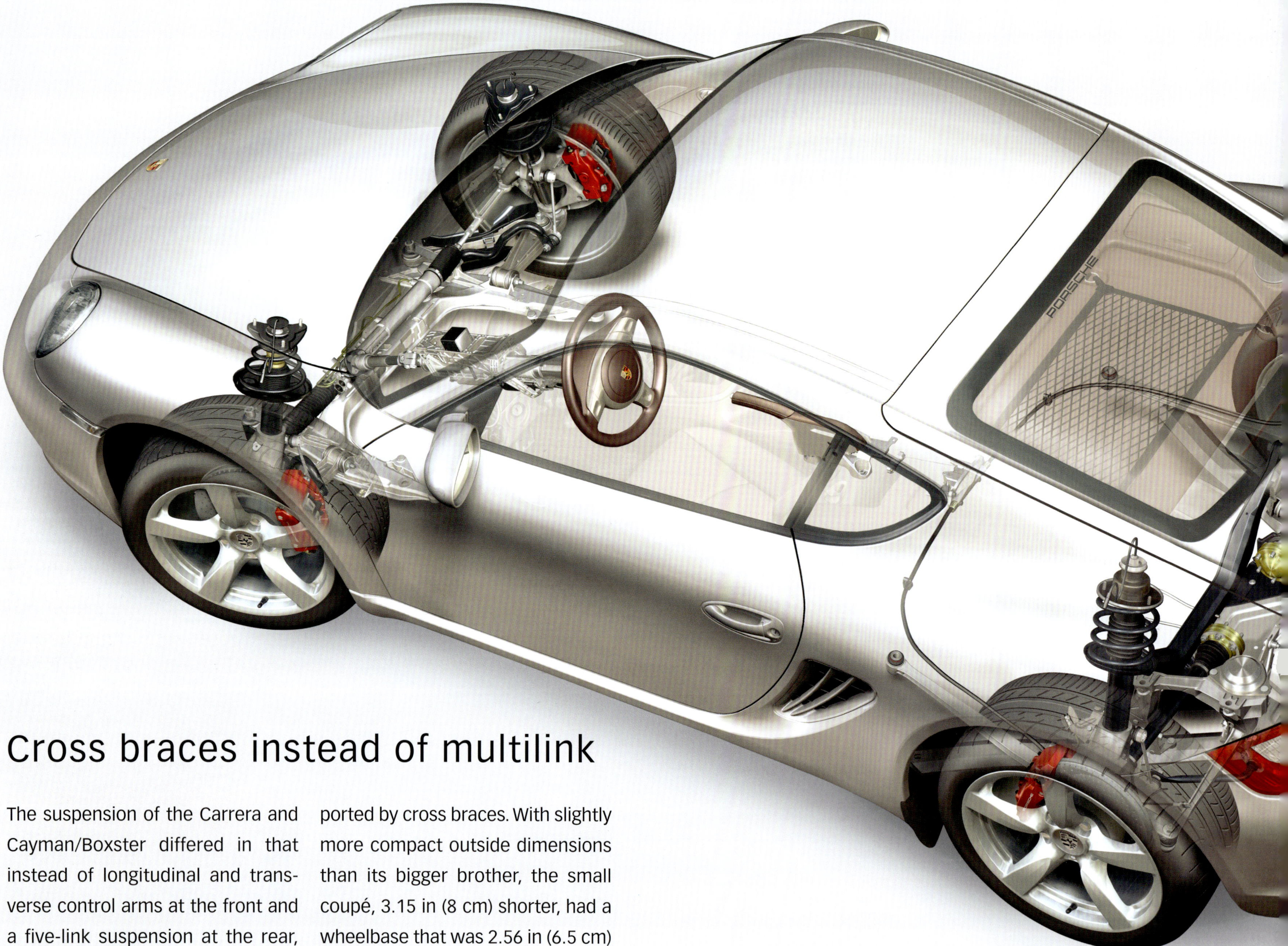

Cross braces instead of multilink

The suspension of the Carrera and Cayman/Boxster differed in that instead of longitudinal and transverse control arms at the front and a five-link suspension at the rear, the MacPherson struts were supported by cross braces. With slightly more compact outside dimensions than its bigger brother, the small coupé, 3.15 in (8 cm) shorter, had a wheelbase that was 2.56 in (6.5 cm) longer.

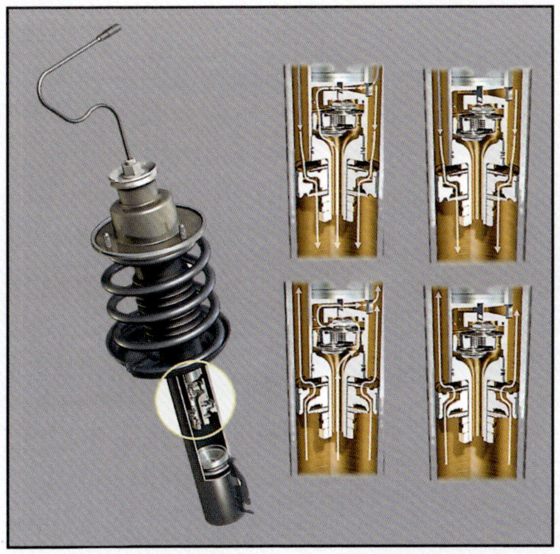

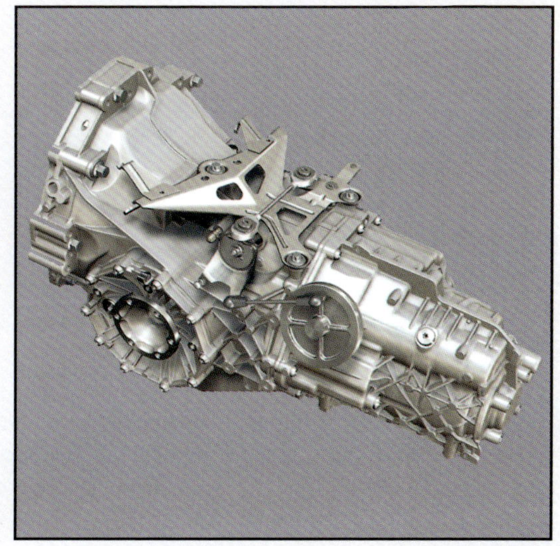

The chassis of the Cayman had MacPherson strut suspension front and rear. The independent front suspension was fitted with coil springs and twin-tube shock-absorbers as well as stabilizers. Optional was the active chassis with PASM (Porsche Active Suspension Management). With it the Coupé was 0.4 in (10 mm) closer to the road surface. The electronically-controlled shock-absorbers had two manually selectable settings: normal and sport. The standard steel brake discs (internally ventilated) could be replaced at extra charge by the Porsche Ceramic Composite Brake (PCCB) with six-piston brake calipers.

The liquid-cooled six-cylinder engine of the Cayman S with a capacity of 206.7 cu in (3,387 cc) developed 295 bhp at 6,250 rpm. As for the 164.0-cu in (2,687-cc) Boxer engine, it was mounted in a mid position in front of the rear axle. It had integrated dry sump lubrication with one pressure pump and two return pumps. The oil tank was in the engine. The thermodynamic cycle was controlled by four valves which were each operated by overhead camshafts. With the variable camshaft timing and valve lift adjustment provided by the VarioCam Plus, the maximum torque remained constant at 250.8 ft lb (340 Nm) from 4,400 to 6,000 rpm.

There were six manually operated speeds in the rear-wheel drive Cayman S. A five-speed Tiptronic S transmission was also available as an option. This automatic gear-shifting was comfortable but it also had disadvantages: the curb weight of the vehicle rose by 88 lb (40 kg) to 3,230 lb (1,465 kg) and acceleration from a standing start to 62 mph (100 km/h) took 6.1 seconds instead of 5.4. Consumption also rose by 5.5 mpg (0.4 liter per 100 km). The remarkable maximum speed of 171 mph (275 km/h) remained unchanged. The Cayman had a five-speed transmission; only the Sport Package, available at extra cos, would include the six-speed transmission.

Porsche Cayenne

The multi-talented sports vehicle

›› As an additional range, Porsche launched the Cayenne on the market In 2002.

It shared its floor pan with the Volkswagen Touareg. The "Carry Over Parts" concept whereby parts of the chassis, the electrics, and the body as well as a few other parts of both makes were shared, was now also used by the Volkswagen-daughter Audi in its Q7. Profitability came more quickly when development costs were spread out in this way.

Porsche built the Cayenne at its plant in Leipzig. The bodies were produced in Bratislava in Slovakia and the engines were made in Zuffenhausen. Success soon followed. The Cayenne quickly became the best-selling Porsche model, and it was particularly successful in the North American market. After starting with a 274.6-cu in (4.5-liter) V8 engine, developing 340 bhp in its normally-aspirated form and 450 bhp as a bi-turbo – it was fitted with a 195.3-cu in (3.2-liter) VR6-engine as an economy version in 2003.

In 2007 Porsche gave the model a facelift by slightly increasing the engine capacity to 219.7 and 292.9 cu in (3.6 and 4.8 liters) , giving them a few more horsepower. The GTS was presented in Frankfurt in 2007 as the fourth variant. The first Porsche diesel car was launched in 2009: the Cayenne with a V6 turbo engine of 195.3 cu in (3.2 liters), designed by Volkswagen. Porsche gave it an additional 30 bhp.

Cayenne Turbo/S

Too good for off-road

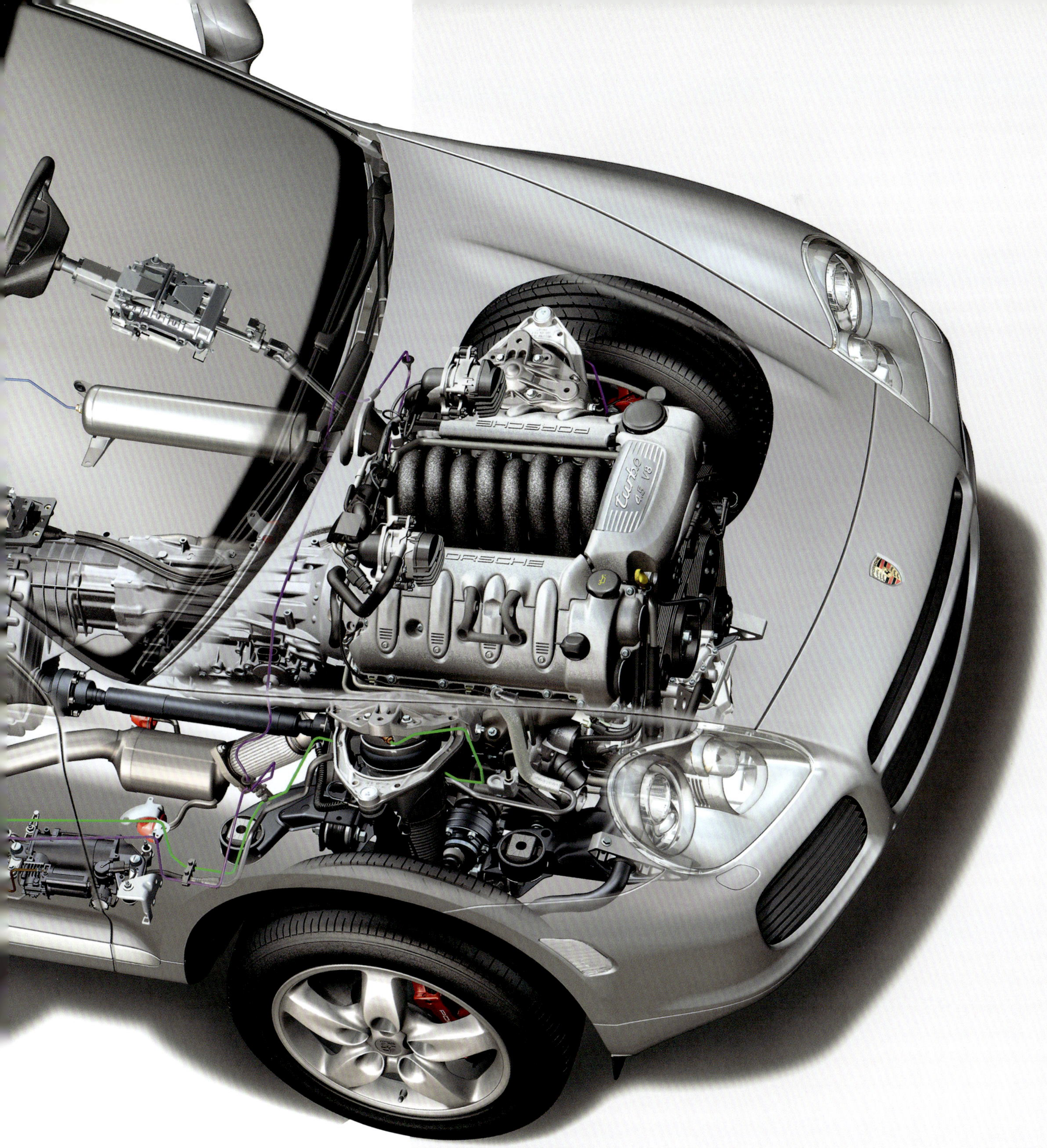

» The leading sports model in the Cayenne program was the Turbo, later the Turbo S. Its 90° V8 engine with four-valve technology, adjustable camshaft, and dry sump lubrication was mounted longitudinally over the front axle. Twin turbochargers in the engine with a compression ration of 9.5:1 provided 450 bhp at 6,000 rpm, and a maximum torque of 479.4 ft lb (650 Nm) that remained even between 2250, and 4750 rpm. In the most powerful version, output was 550 bhp with a torque of 553.2 ft lb (750 Nm). The car with the self-supporting steel body providing enough space for five passengers had a six-speed Tiptronic transmission but later the model also became available with a manual six-speed transmission. Permanent all-wheel drive and an electronic self-locking differential ensured a smooth drive even on impassable terrain so long as the correct tires had been fitted. The power was divided so that 38% went to the front wheels and 62% to the rear wheels.

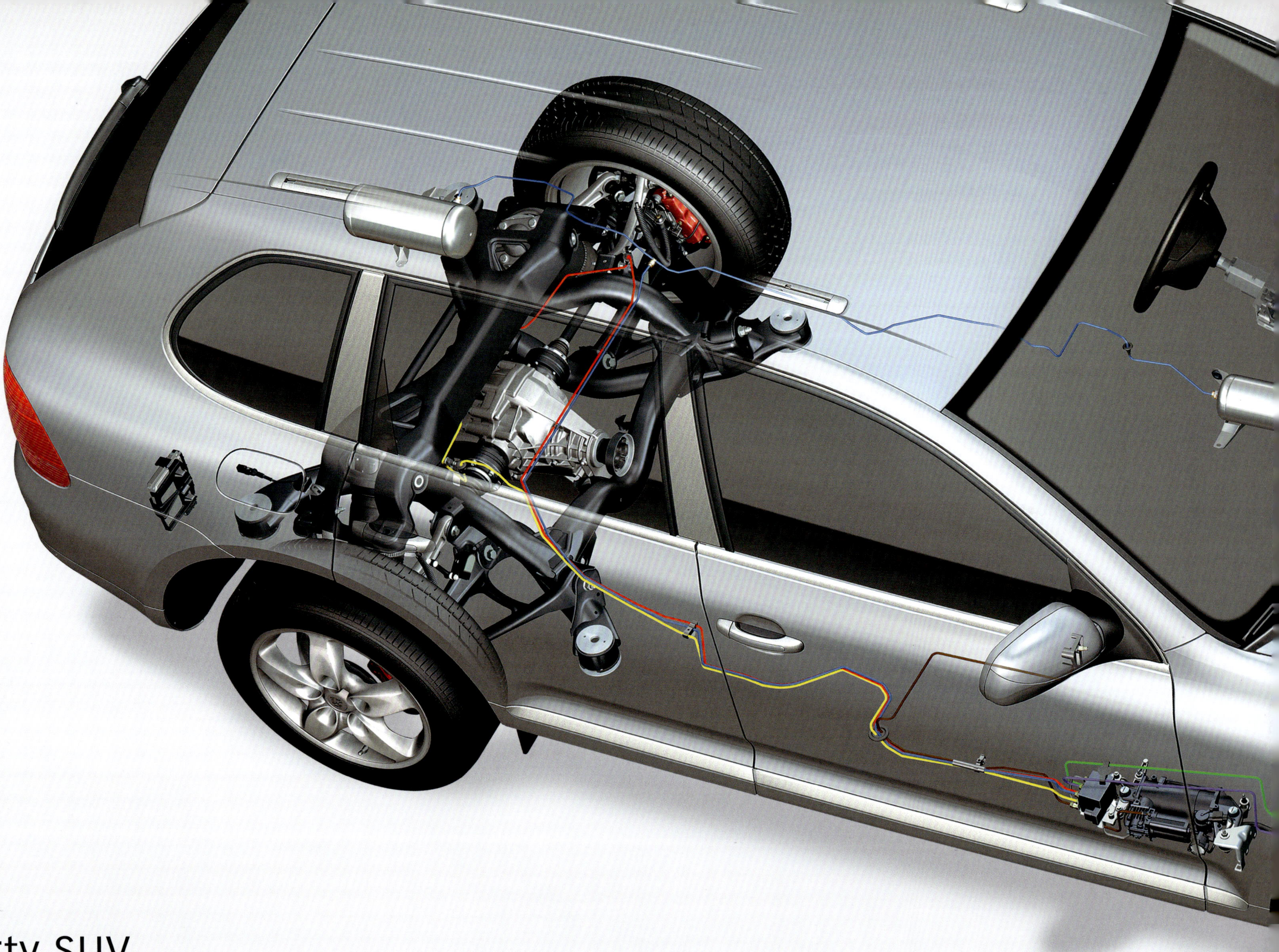

A sporty SUV

The front suspension consisted of double wishbones, coil springs, and stabilizer. Electronically controlled air suspension, available for other models as an option, was standard on the Turbo. The curb weight of 5,192 lb (2,355 kg) and a speed of up to 174 mph (280 km/h) in the final version of the vehicle required a special braking system: ABS and automatic differential braking were included, with four-piston brake calipers on the rear axle, six-piston calipers at the front, and ventilated discs.

Double overhead camshafts per cylinder bank and four valves per cylinder controlled the engine, supplied by two turbochargers with intercooling. In 2007 the DFI Direct Fuel Injection system replaced the Bosch Motronic. The designers decided to drive the camshaft with heavy chains. The adjustable camshaft made the useful torque speed range wider. Maximum torque was reached at a modest 2,250 rpm.

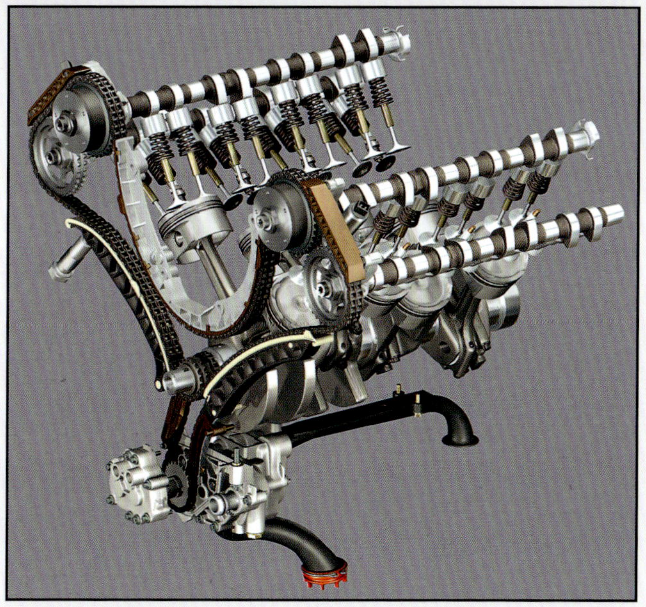

170-171 A cutaway illustration made from 3D data, used by Technical Art for the first time on the Cayenne. The car in cutaway is shown on the road in realistic surroundings (3D background photo technique).

S NR 3028
D
PORSCHE

Cayenne S

Built strong

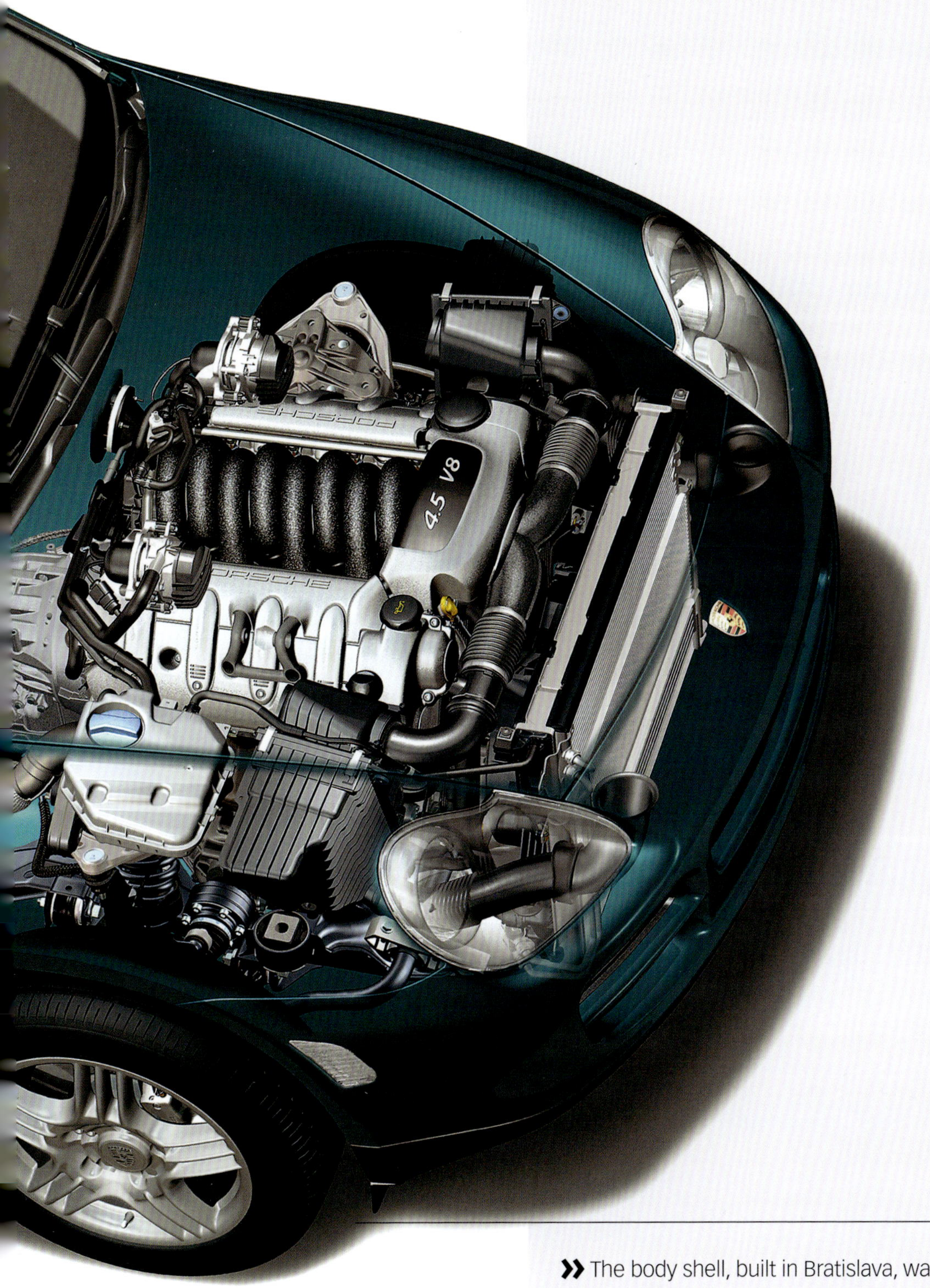

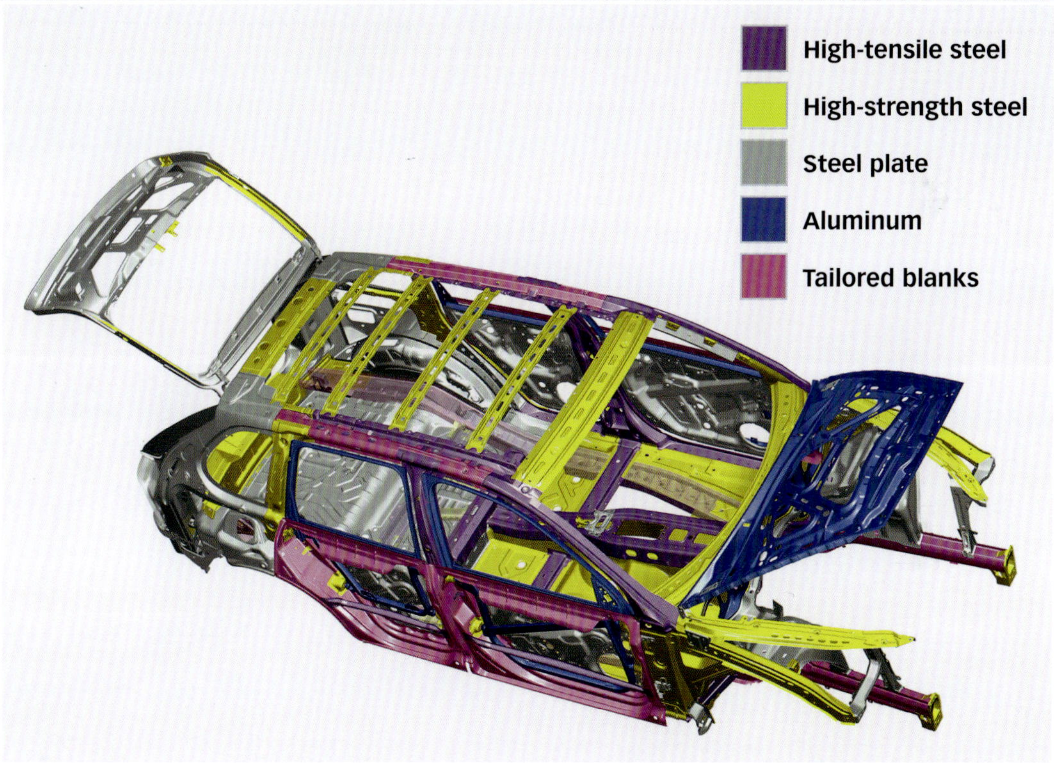

» The body shell, built in Bratislava, was made of several materials. High-tensile steel ensured maximum crash safety. But this special material is expensive, so ordinary high-strength steel parts or simple steel panels were used in other places. On the other hand, parts with elaborate shapes are were produced as "tailored blanks." The hood was aluminum, a light metal.

The V8 engine without turbocharging was designated M48/00. Like its turbocharged counterpart, it had a capacity of 275.28 cu in (4,511 cc). In the model year 2007 the bore and stroke of 3.66 x 3.27 in (93 x 83 mm) were increased to 3.78 x 3.27 in (96 x 83 mm), bringing the capacity to 293.3 cu in (4,806 cc). The normally-aspirated engine had a higher compression ratio (11.5:1), and developed 340 bhp at 6,000 rpm while the maximum torque was 309.8 ft lb (420 Nm) at 2,500 to 5,500 rpm. With the increase in engine capacity, the specifications: compression ratio 12.5:1, 385 bhp at 6200 rpm, and 368.8 ft lb (500 Nm) at 3,500 rpm.

In the permanent all-wheel drive of the Cayenne range, 38% of the power went to the front wheels and 62% to the rear. An electronic differential distributed the driving torque and evened out the differences in rotational speed between the front and rear axles. The rear axle differential shown here evened out the differences in rotational speed between the two rear wheels. As a result, propulsion continued even if some wheels had lost contact with the surface of the terrain.

Cayenne VR6

Finer than a VW

» In 2003 the Cayenne VR6 was launched on the market at a starting price of 50,000 Euros. It was about 13,000 Euros less than the Cayenne S with a V8 engine. The VR6 engine had a very narrow angle between cylinder banks of only 15°. Porsche had revised this engine, originating from Volkswagen, and given it an extra 30 bhp; as a result the Cayenne was could accelerate from a standing start to 62 mph (100 km/h) in nine seconds and had a maximum speed of 133 mph (214 km/h).

In 2007 the engine capacity rose to 219.6 cu in (3,598 cc) while the horsepower increased to 290 bhp, and accordingly the performance improved; it could now accelerate from a standing start to 62 mph (100 km/h) in eight seconds and the maximum speed was 141 mph (227 km/h).

The multi-link rear suspension had standard coil springs, and stabilizer. To order and at extra cost, the VR6 was also available with electronically controlled pneumatic suspension. The internally ventilated brake discs were 13.0 in (330 mm) in diameter and four-piston fixed calipers came into action with great force when needed.

Clearly visible is the position where the rear axle differential with the drive shafts is mounted.

In principle the drive train of all Cayenne variants was mounted in the same way as a permanent all-wheel drive.

The six-speed Tiptronic, or manual transmission if requested, conveyed the power to the rear wheels and via an auxiliary drive to the front axle. For challenging terrains there was an intermediate transmission. Electronics also played a part here: the self-locking differential and wheel slip traction control were computer-controlled.

The throttle valve housing of the compact VR6 engine looked like a little work of art. A vital control component, it was mounted between the air filter and the intake manifold. In the Cayenne petrol models, regardless of the number of cylinders, the fuel mixture was controlled by a Bosch DME Motronic 7.1.1. The DFI direct fuel injection system introduced by Porsche in 2007 gave large savings in fuel consumption. The body's aerodynamics were also improved when the model was upgraded.

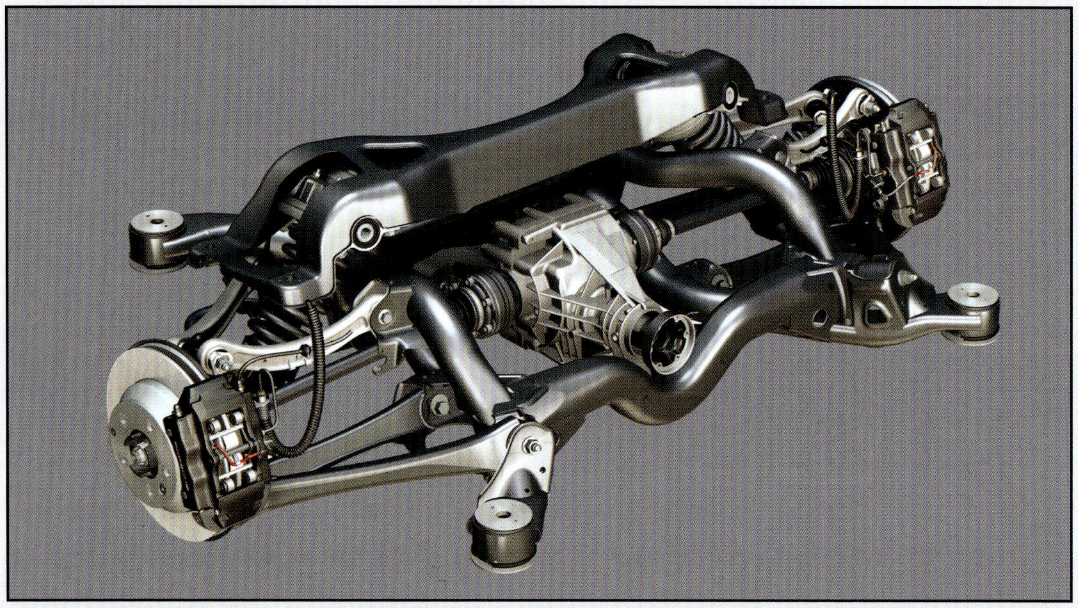

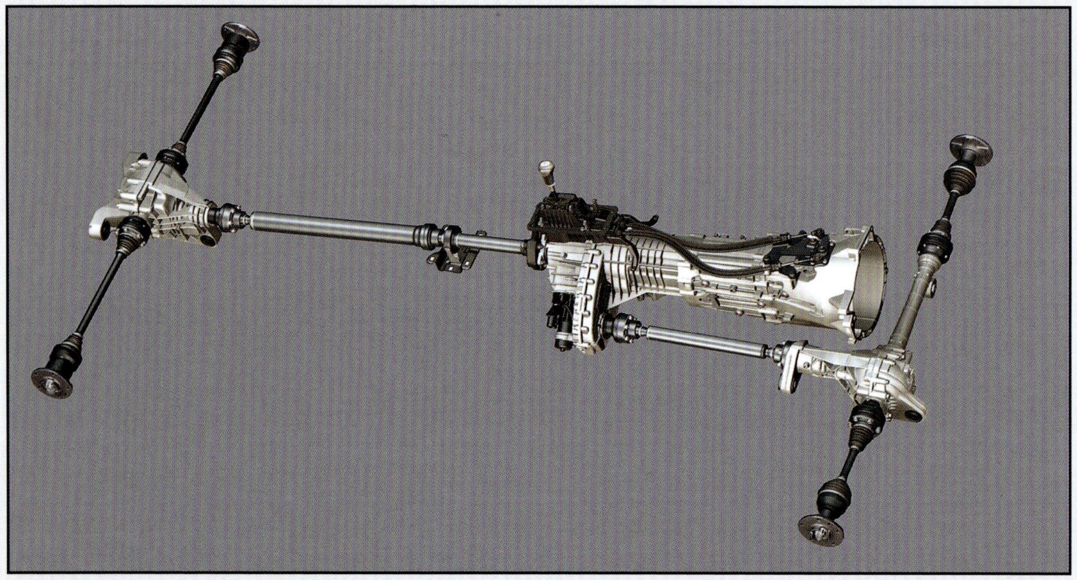

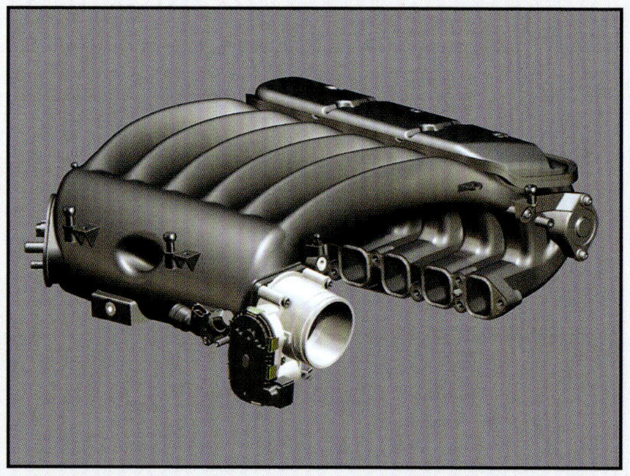

S GO 1120

Panamera GTS

Sports car with four doors

» **In January 2012 Porsche** launched the purist Panamera GTS as a sports version of the four-seater. The name is says it all: Porsche GTS, meaning "Gran Turismo Sport," and it promises the most extraordinary performance since the legendary 904 Carrera GTS. The sportiest of all Panameras so far produced, it conveys this character to its occupants and the outside world through its distinctive sound. The body has been lowered by 0.4 in (10 mm) and the suspension is firmer, changes that have given this additional model variant great nimbleness and sports handling qualities.

More power, more powerful brakes, and the corresponding sports chassis with pneumatic suspension and PASM were the most important technical modifications made to create a Panamera that was also suitable for circuit racing, but without reducing its suitability for daily use. The appearance of the new GTS clearly reveals its personality in several distinctive design features on the front, sides and rear, as well as with contrasting black highlights. The interior too has been tailored to match the sports personality of the model with specially adapted sports seats, Sport Design steering wheel with shift paddles, and an exclusive leather interior with Alcantara surfaces.

The Panamera GTS power unit

›› The power unit of the new Panamera GTS is a modified 292.9 cu in (4.8-liter) naturally aspirated V8 engine developing 430 bhp at 6,700 rpm and therefore 30 bhp more than the engine mounted on its sister-model the Panamera S/4S. The maximum torque also increased, from 368.8 to 383.5 ft lb (500 to 520 Nm). The Porsche twin clutch transmission (PTM/TCT) conveys the power to the all-wheel drive (PTM) at the mere push of a button, without interrupting the tractive power, by means of the standard Sport-Chrono Package. The interaction between engine, transmission, and all-wheel drive enables the car to reach 62 mph (100 km/h) in 4.5 seconds and then accelerate to its maximum speed of 179 mph (288 km/h). With a consumption of 21.6 mpg (10.9 liters per 100 km) the Panamera GTS remains as efficient as all other Porsche models. The standard values for acceleration and maximum speed do not do justice to the characteristics of this sports limousine. Its chassis in particular makes it not only a brilliant four-seater sports car for everyday use but also perfectly suited to compete on race circuits: the variable pneumatic suspension and Porsche Active Suspension Management (PASM) are standard. The chassis can be adjusted according to driving conditions by means of the ride-height control system, vertical adjustment, variable spring characteristics, and electrically adjustable shock-absorbers. The 19-inch tires measuring 255/45 on the front axle and 285/40 on the rear axle provide the necessary road grip. The Panamera GTS is fitted with the brilliant brake system of the Turbo model.

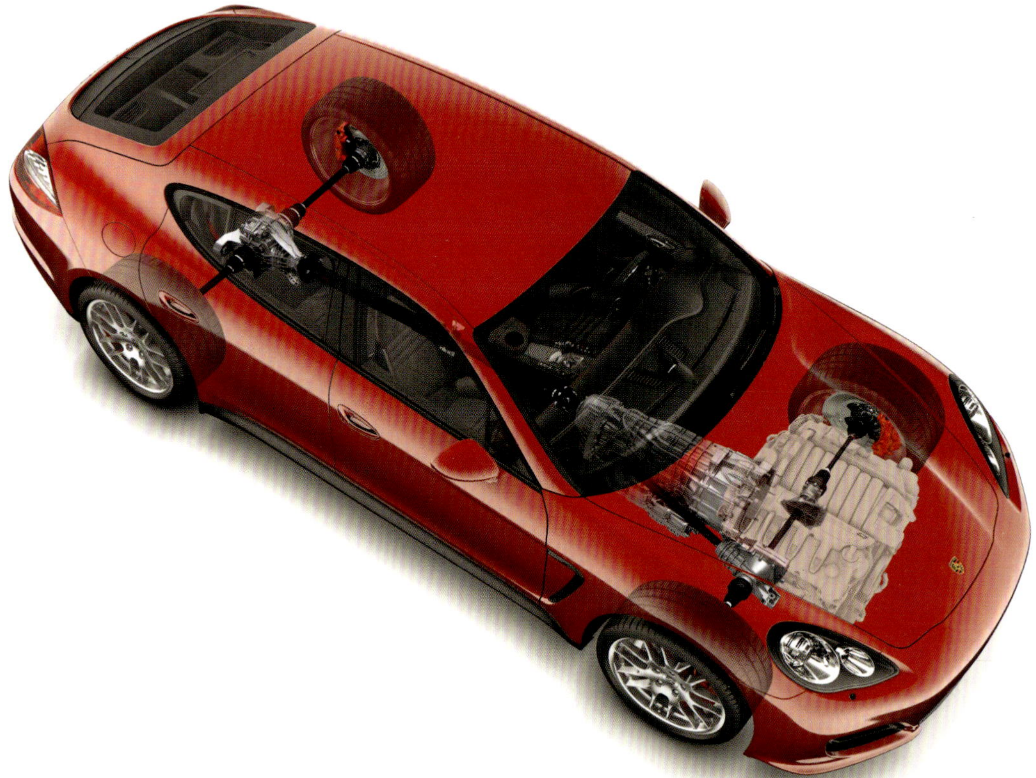

Panamera GTS

03
03
E/4
E/4
Informationen
Layout
Team 4 Rädermontage
08

Package solution

» **A Porsche** has very distinctive features and the body houses a very specific technology. Both aspects have been extensively illustrated and described in the previous pages of this book.

But when developing a car, it is neither the technology nor the design that is the deciding factor, because the whole package must work as a whole. Both aspects become merged in the package. Here you can read the actual report by the man who is responsible for the entire product range of the Stuttgart manufacturer. He describes how a typical Porsche comes into being and how a successful car is created from the parameters of the specification sheets and the design ideas.

Car concept and package

by August Achleitner

General

›› The car concept is the constructive outline of an idea for a car in which the feasibility of the idea is tested by compiling all the actual geometric specifications and the relevant components that define the concept of the vehicle. This concept of the car is then developed further into the car package that through the specification of all the research and components will ensure its geometric viability before putting it into production.

But before we come to the subject of car concept projects, a vehicle profile must be established in which the actual target objectives of the new product will be described. These are as follows:

- Vehicle type and class (e.g. 2+2 seater, rear-engine sports car)
- Body (e.g. coupé or convertible)
- Interior arrangement and useful volumes (e.g. 2+2 seating, luggage capacity, fuel tank capacity)
- External dimensions (approximate dimensions)
- Engine (type and power)
- Drive train (transmission and drive, for instance rear-drive and all-wheel drive)
- Body concept (e.g. standard steel construction or aluminum space frame)
- Chassis, axle concepts, possible wheel sizes
- Weight goals
- Aerodynamic goals
- Driving performance and fuel consumption targets
- Intended markets
- Market, customer or segment-specific requirements (including legislation) (e.g. target retail price)
- Characteristics (e.g. USP – unique selling proposition).

August Achleitner is responsible for the "package" of every new Porsche model.

The creation of a vehicle profile from which, after further itemization, the specifications will later be established, is usually simpler when defining a successor model than in the case of a new model series. When dealing with a successor model the "to do" list is known from the predecessor, such as conformity with future legislation, the achievement of particular consumption and emission targets and the avoidance or reduction of not completely satisfactory features of the previous model (for instance, reducing the cost of ownership). In a fundamentally new range of models, drawing up the vehicle profile for approval by the board of management is a protracted business because the exact positioning of the new vehicle in the competitive environment must be defined after intensive research in order not to jeopardize the vehicle's future success on the market.

No construction work takes place until the vehicle profile has been established; styling sketches to provide a visualization of the idea of the car are developed in parallel where required. The actual start of the development work comes with the approval of the vehicle profile. At this point the project management team – in the case of Porsche, the model range organization – takes over the running of the project and with the collaboration of all the departments it fleshes out the vehicle profile to a 1,000-page list of detailed specifications. This list contains all the requirements expected from this vehicle and these are then precisely described in every details. This will provide a guideline during the development of the project and throughout the construction process of the vehicle

At the same time as the approval of the vehicle profile, concept work starts in the department of vehicle concepts and packages. The new model starts life as a purely technical skeleton and is then built up "from the inside to the outside." In other words, taking into account the chosen vehicle concept (for instance, sports car, rear engine) and also the target parameters, the vehicle width, height, and length, seat position and ergonomic passenger compartment dimensions (for instance, the position of the pedals, steering system, seat travel adjustment) are developed. This is accompanied by the necessary studies in the area of the vehicle's underbody (drive train, cables, body structures). With the arrangement of the usable volume and of the essential mass-determining components (for instance, the engine, transmission, fuel tank, radiator, trunk compartment volume, exhaust mufflers, battery, and wheels), the new vehicle comes to life as a general technical structure that can now undergo the first simulated tests.

The vehicle concept is developed further as a flowing transition into the "vehicle package" that now forms the basis for the first calculations that will be used to test the establishment of weight targets, axle weight distribution, driving dynamics, and consumption targets.

The vehicle concept sketches can now present things visually: for instance, as four vehicle occupants who, sitting in the vehicle, are surrounded by all the essential technical components, possibly even including luggage in an imaginary trunk, but without any bodywork, so that visually it is a little like a composition in a go-kart. What is still missing is a "skin" of any kind, any styling of either the inside or the outside of the outer shape.

The creation of the first vehicle concept sketches provides the stylists with these data, now of course no longer in the form of drawings but as CAD data records that define the technical limits. Indeed, the concrete design work on a project does not start exactly in parallel with the technical concept work but is slightly displaced in time. For the designers, when creating the styling surface areas, a "violation" of the technical surfaces is taboo; they can position their surfaces only up to the technical limits but they are not allowed to penetrate them because otherwise the technical feasibility could not be guaranteed.

This is how things work in theory. In reality there are often conflicts of goals between the technical parameters and styling, when designers ideally would like a different shape from that made possible by the technical aspects. The result is an iterative process in which the best compromise between technical practicability and attractive design is sought . The use of CAD (Computer Aided Design) and CAS (Computer Aided Styling) technology, combined with simultaneous photo-realistic representation, allows a much quicker assessment of the solutions proposed than was possible in the past.

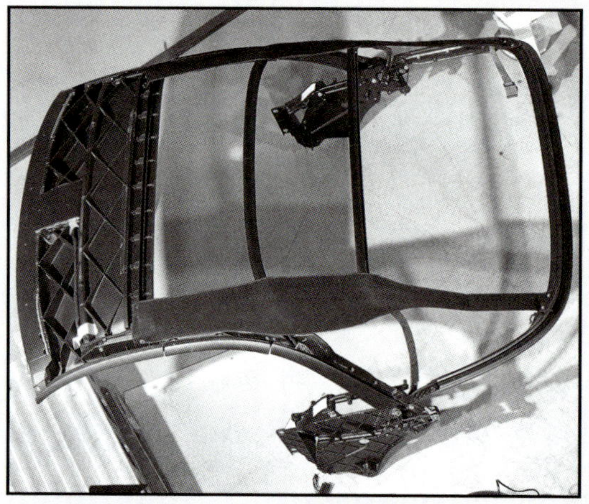

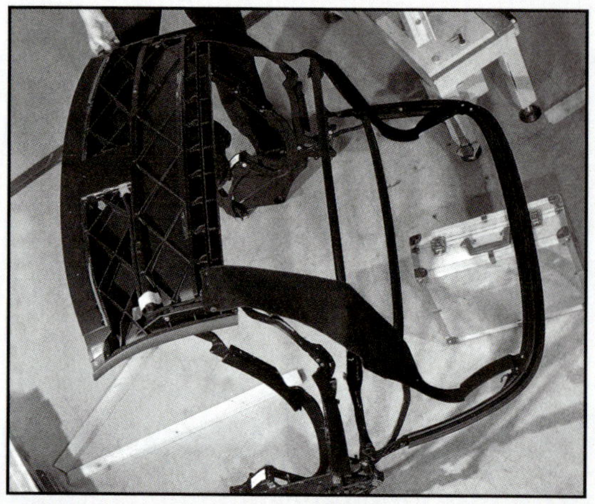

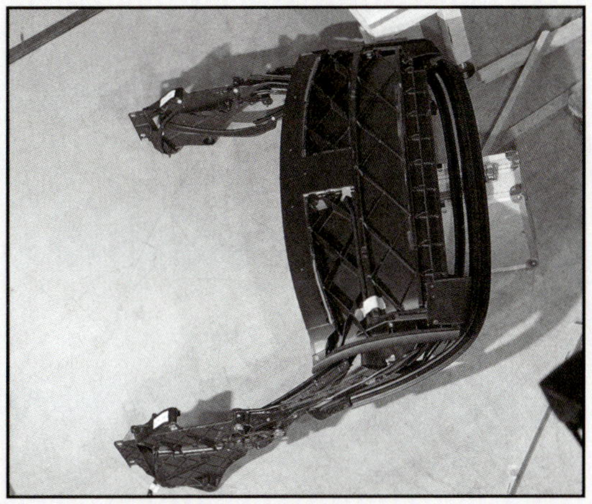

Porsche specification

The Porsche 911 – an insight into the concept of success

›› The 911 is the most successful sports car in the world with a unique rear-engine concept, a 2+2 interior, and a typical styling line, conceptually unchanged for over 45 years.

The rear-engine design has specific advantages, especially its unique mechanical traction from the rear weight bias arising from the fact that the engine is behind the rear axle, so that the rear axle load is between 58%, and 62%, depending on the variant. As with a mid-engine concept, the mechanical path between the crankshaft and the transmission output through the rear wheels is extremely short, so it is mechanically efficient, and also slightly less subject to friction. Correspondingly, the arrangement requires the positioning of the fuel tank in the front of the car, crash-protected behind the front axle, but at the same time limited in maximum volume. The concept vehicle was on the one hand an excellent sports car for everyday use with a good amount of room, and on the other hand it demonstrated conceptual advantages as an extreme competition car. The 911 is the most successful racing car of all time and over 200 racing cars are produced each year.

Increasingly stringent emission regulations, increasing engine performance, and reduced consumption by consumers in the 1990s made it necessary for the 911 type 996 to switch to water-cooled 4-valve engines. Each cooling radiator was mounted in front of the front wheels to achieve a favorable air flow, and at the same time to avoid obstructing the space between the front wheels, so that the trunk remained of acceptable size.

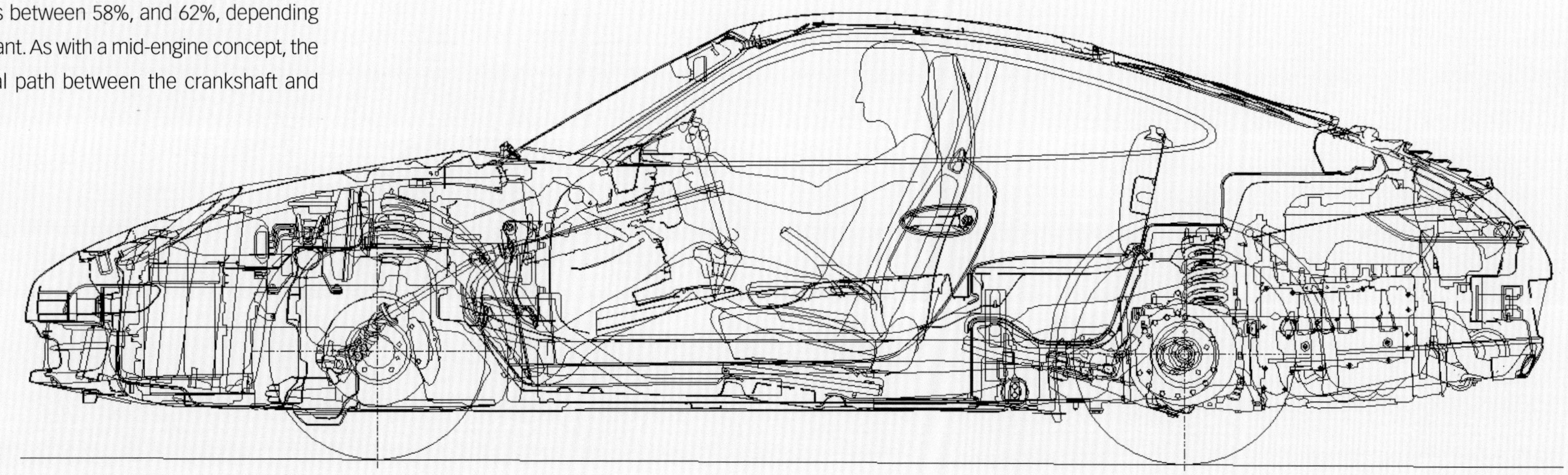

The excellent handling characteristics of the 911 – a result of the overall concept taking winter into account – was further increased by all-wheel drive models. The all-wheel drive Carrera 4 and Turbo models are now the most successful models in the 911 range. In these variants the drive train of the rear-drive base model is supplemented by the so-called hang-on front axle, the drive force being transmitted to the front through an electronically-controlled multi-plate clutch and longitudinal coupling. The low-profile six-cylinder boxer engine in the rear is ideally suited for this position, and together with the turbo version of this engine, outputs ranging from 340 bhp to over 600 bhp are obtained, supplemented in motor sport situations by specially optimized assemblies.

Specifically, the 911 models have a performance that gives the slowest variant a top speed of over 174 mph (280 km/h), and up to 196 mph (315 km/h) in the fastest versions. This requires vehicle aerodynamics that will on the one hand have low air resistance (drag coefficient) to ensure low fuel consumption; on the other hand, they must create a low lift coefficient to ensure safe driving behavior at very high speeds. These lift coefficients (Czf – front axle and Czr – rear axle) are close to 0 in the "normal" 911 models, and significantly below 0, that is, downward, in the high-powered turbo and GT models. Achieving this is possible only with aerodynamic aids, such as a retractable rear spoiler, and by meticulous optimization of the vehicle in a wind tunnel, including the underbody, wheel arches etc. In the vehicle the appropriate components and techniques will be considered (e.g. actuated rear spoiler or air ducts). Aerodynamic requirements often exacerbate the already existing trade-offs between the package and styling.

Four-door sports sedan and an SUV

» From a purely conceptual viewpoint, the Cayenne and Panamera ranges represent the "normal" characteristics of an SUV and a sports sedan respectively. Both ranges have a front engine mounted longitudinally with transmission mounted directly on it and a drive shaft to the rear axle. As well as rear-wheel drive, the Panamera is also available with various engines in an all-wheel drive version. The all-wheel drive in the Panamera and Cayenne is conveyed to the front axle via an electronically controlled longitudinal coupling. But these front-engine ranges also have distinctive features in their appearance. The body of the Panamera is unusually low compared to the competition, emphasizing its closeness to the sports car ranges and conceptually representing a unique selling point. The floor platform that the Cayenne shares with the VW Touareg cars and (in modified form) in the Audi Q7 represents a significant increase in the number of components used in common, and the consequent spreading of one-off development costs will have had a beneficial economic effect. The basic layout of the Cayenne has the same ergonomic features as the 911, for instance the zero offset between the middle of the driver's seat and the steering column, as well as the position of the steering wheel parallel to the vehicle (not skewed).

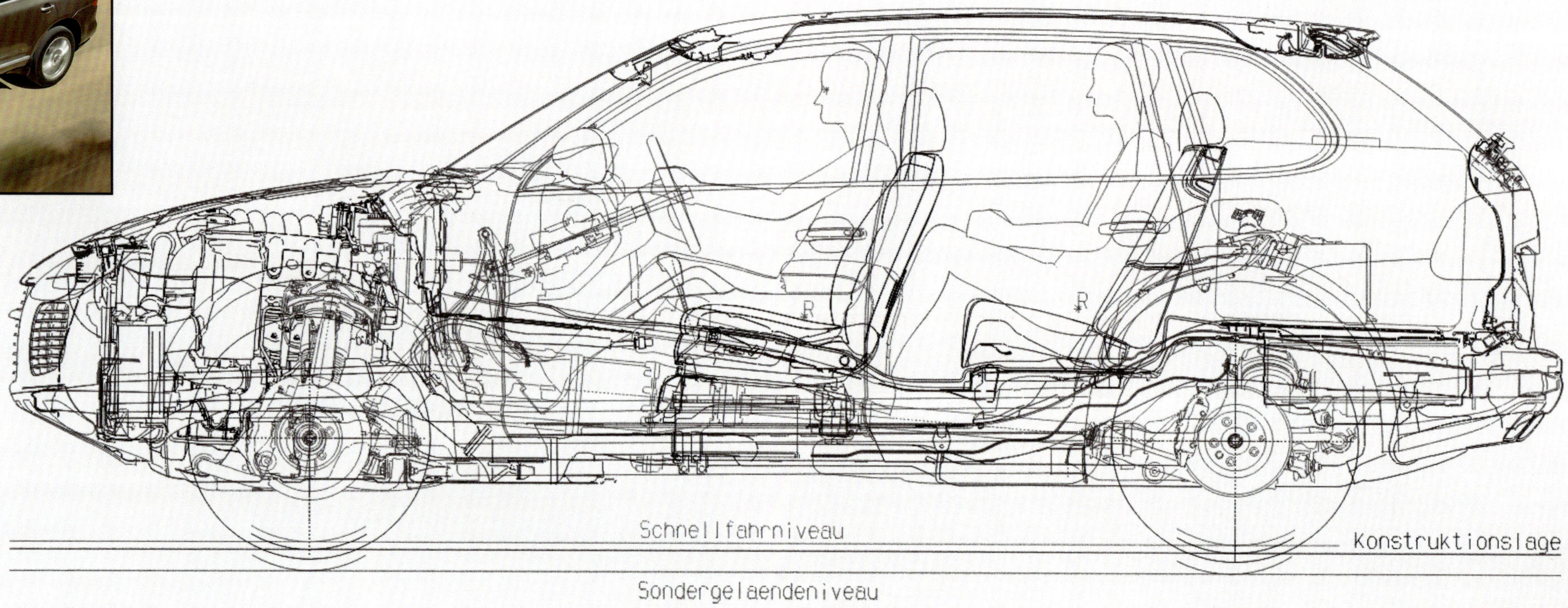

A pure two-seater

» With Porsche's sports cars too there have been platform concepts, families of parts in common, and technical concepts to optimize the economics of project development. An example of this concept of using the same component with variations is the underside of the 911 in the mid-engine Boxster sports car (as a Roadster, and the Cayman as a coupé). The floor platforms of the rear-drive 911 and the Boxster up to the "heel-plate" behind the front seats are identical, and they are individualized only by specific add-on parts in the interior and exterior. In the Boxster the transmission is located behind the engine, and the engine itself (apart from its capacity and certain performance-limiting components) is largely the same as that of the Carreras. Only the rear section (from the heel plate) with the rear axle is unique to the Boxster specification. The limited sports car market could be enlarged by the addition of a range of mid-engine cars below the price of the 911. This enabled the capital costs to be distributed over a larger total volume, resulting in lower unit costs for the shared components. With it, the creation of a car priced below of the 911 series was achieved, as well as cost optimization of the 911 series.

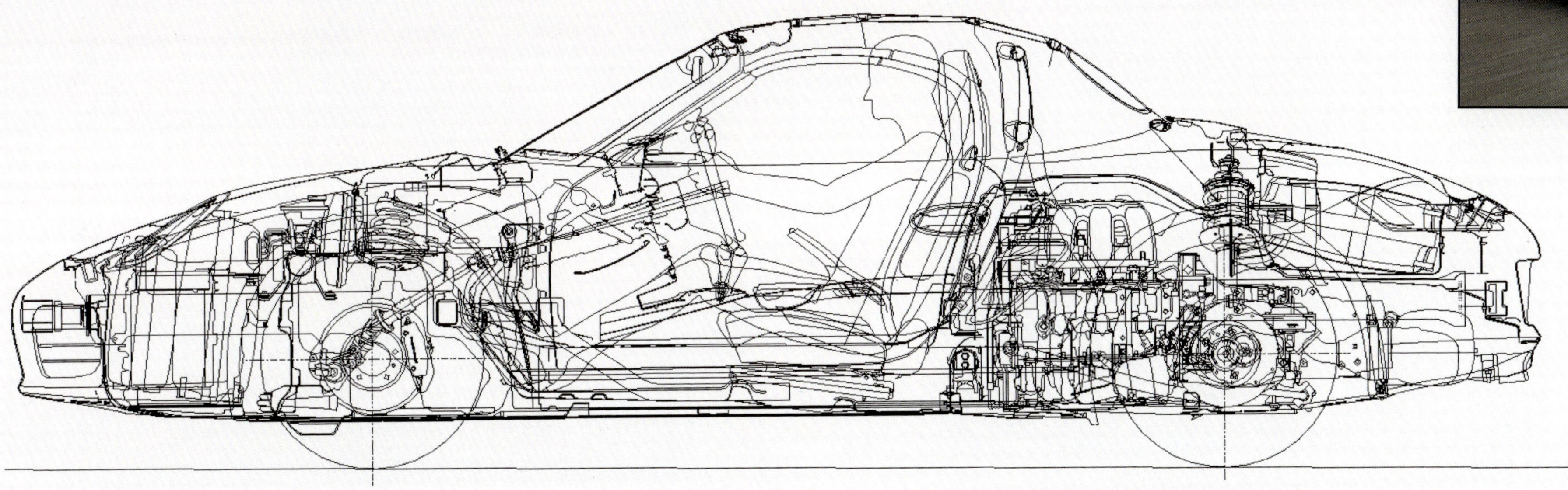

turbo

Bibliography

Lothar Boschen, Jürgen Barth,
Das große Buch der Porsche-Typen,
Motorbuch Verlag, Stuttgart, 1989

Jörg Austen,
Porsche 911 Rallye- und Rennsportwagen. Die technische Dokumentation,
Motorbuch Verlag, Stuttgart, 2005

Peter Schneider,
Typenkompass Porsche Renn- und Rennsportwagen seit 1948,
Motorbuch Verlag Stuttgart, 2003

Jörg Austen,
Porsche 924 – 944 – 968. Die technische Dokumentation der Transaxle-Modelle,
Motorbuch Verlag, Stuttgart, 2003

Jörg Austen,
Porsche 928. Die technische Dokumentation des Transaxle-Achtzylinders,
Motorbuch Verlag, Stuttgart, 2004

Marc Bongers,
Porsche. Serienfahrzeuge und Sportwagen seit 1948,
Motorbuch Verlag, Stuttgart, 2004

David Hemmings,
Porsche 928. Praxisratgeber Klassikerkauf,
Heel Verlag, Königswinter, 2005

Peter Morgan,
Das Original: Porsche 924/944/968,
Heel Verlag, Königswinter, 2006

Halwart Schrader,
Porsche 924/944/968. Schrader Typen-Chronik,
Motorbuch Verlag, Stuttgart, 2006

Jan-Hendrik Muche,
Porsche 924 und 944. Mit vier Zylindern zum Erfolg,
Heel-Verlag, Königswinter, 2002

Börry Lauenstein,
Porsche Typenkunde. Alle Serienmodelle seit 1950,
Delius Klasing Verlag, Bielefeld, 2005

Werner Oswald,
Deutsche Autos 1945 – 1990,
Motorbuch Verlag, Stuttgart, 2001

Elmar Brümmer, Jutta Deiss, Reiner Schloz, Peter Vann,
Porsche Carrera GT,
Motorbuch Verlag, Stuttgart, 2003

Elmar Brümmer, Jutta Deiss, Reiner Schloz,
Porsche Cayman. Jagdfieber,
Delius Klasing Verlag, Bielefeld, 2005

Thomas Agethen, Sigmund Walter,
Typenkompass Porsche, Personenwagen seit 1948,
Motorbuch Verlag, Stuttgart 2003

Tobias Aichele,
Porsche 91 – Forever young,
Motorbuch Verlag, Stuttgart, 2004

Paul Frère,
Die Porsche 911 Story,
Motorbuch Verlag, Stuttgart, 2002

Dr. Thomas Gruber, Dr. Georg Konradsheim,
Carrera RS,
T.A.G. Verlag, Wien,1992

Achim Kubiak,
Faszination 911, Die Typologie des Porsche 911,
Delius Klasing Verlag, Bielefeld, 2004

Randy Leffingwell,
Porsche 911– Perfektion und Design,
Heel Verlag, Königswinter, 2007

Nicky Wright,
Porsche, Traumauto made in Germany,
Orbis Verlag, Munich, 1993

Acknowledgments

The author thanks the Porsche company for providing illustration material from its archives, and the always excellent press relations department: Anton Hunger, longtime Head of Public Relations, Hans-Gerd Bode (Head of Product and Technology Press), Hermann-Josef Stappen (Head of Technology Press), Achim Stejskal (Head of the Porsche Museum), and Dieter Landsberg (Head of Historic Archives) for their cooperation, and the rapid communication of requests for information to departments.

Stefan Buescher (former Head of Marketing and Communication), Bjoern Ewald (Global Training), Dr. Ulrich Lutz (Technology After-Sales), and Jens Thomas (Consulting) have provided invaluable and helpful support in preparing the drawings.

We also thank the former Porsche designer Harm Laagay for his helpful suggestions.

Authors

Lars Döhmann, is a German motor journalist. For decades his articles have been published in magazines; a freelance since 1995, he has also worked for the German newspaper *Frankfurter Allgemeine Zeitung* and the Oldtimer Catalog. In addition, Döhmann is organizer of the "2000 Kilometers through Germany" Oldtimer Rally. In his free time he looks after a 1979 Porsche 911 SC 3.0 Targa.

Norbert Schäfer set up his first studio in 1965. Even then his illustrations demonstrated his love of detail and, most of all, his "X-ray vision," the ability to visualize that is so indispensable for technical illustration, together with his skills at realizing spatial arrangements and perspectives from blueprints and photographs. In 1975, together with his business partner and friend of many years, Ludwig Eberl (died 1995), who shared his passion for technical subjects, he founded Technical Art, and this was the beginning of the collaboration with Porsche. Today he still works there in his capacity as owner-manager.

Cover: Porsche 993 Turbo.
Backcover: Porsche 997 Carrera.

Product management: Martin Distler
Layout concept: Thomas Uhlig
Layout: Gaby Herbrecht
Editing: Ute König
Cover design: Thomas Uhlig
Repro: Cromika s.a.s.
Production: Thomas Fischer

All information in this work has been carefully researched by the authors and brought up to date and tested by the publisher. The information is beleived to be accurate but no liability can be accepted. We are grateful for any comments and suggestions. Please send them to:

WHITE STAR PUBLISHERS

WS White Star Publishers® is a registered trademark property of Edizioni White Star s.r.l.

Via M. Germano, 10 - 13100 Vercelli, Italy
www.whitestar.it

Translation, editing and layout: Rosetta Translations SARL

ISBN 978-88-544-0678-0
1 2 3 4 5 6 16 15 14 13 12

Printed in China